PHOTOSYNTHESIS: PHYSIOLOGY AND METABOLISM

Edited by
Cajumban, Alyssa A.

ArclerPress

Photosynthesis: Physiology and Metabolism

Edited by Cajumban, Alyssa A.

Edition 2018

a
ArclerPress

Published by
Arcler Press LLC
708 3rd Avenue, 6th Floor
New York
NY 10017
United States of America
www.arclerpress.com

ISBN: 978-1-68094-541-6 (Hardback)

Printed in United States of America

Library of Congress Control Number: 2017933112

About the Editor

Cajumban, Alyssa A.

Alyssa Cajumban graduated with a degree in Biology at the University of the Philippines - Manila. In the broad field of biological sciences, her interests focus on microbiology and plant pathology. Some of her works include researching and writing scientific papers regarding fungi, diseases of agricultural crops, and post-harvest techniques. She also works as an editorial assistant and writer for a scientific journal and textbook publication

Table of Contents

3 THE ROLE OF CHLOROPLAST

4 LIGHT-DEPENDENT REACTIONS

5 PHOTODISSOCIATION IN PHOTOSYNTHESIS

6 CARBON FIXATION: THE C2, C3, C4 PATHWAYS

7 CRASSULACEAN ACID METABOLISM CAM PHOTOSYNTHESIS

8 RUBISCO PROCESS

9 EMERSON & WARBURG EFFECT

10 PHOTOSYNTHETIC PIGMENTS

List of Abbreviations

ADP	Adenosine Diphosphate
ATP	Adenosine Triphosphate
ALA	Aminolevulinic Acid
BASF	Badische Anilin- & Soda-Fabrik
BIRD	Blackbody Infrared Radiative Dissociation
EET	Electronic Energy Transfer
EPR	Electron Paramagnetic Resonance
ETC	Electron Transport Chain
EPGase	Enolase-Phosphglycerase
CCM	Carbon-Concentrating Mechanism
CO_2	Carbon Dioxide
CAM	Crassulacean acid metabolism
DCA	Dichloroacetic Acid
LSC	Long Single Copy Section
MODIS	Moderate Resolution Imaging Spectroradiometer
MPD	Multiple photon dissociation
OEC	Oxygen Evolving Complex
PEPC	Phosphoenolpyruvate Carboxylase
PEP	Phosphoenolpyruvate
PGAL	Phosphoglyceraldehyde
SSC	Short Single Copy Section
SWCNT	Single Wall Carbon Nanotube

Preface

Photosynthesis is the process used by plants, algae and certain bacteria to harness energy from sunlight into chemical energy. It is one example of how people and plants are dependent on each other in sustaining life. Photosynthesis happens when water is absorbed by the roots of green plants and is carried to the leaves by the xylem, and carbon dioxide is obtained from air that enters the leaves through the stomata and diffuses to the cells containing chlorophyll. The green pigment chlorophyll is uniquely capable of converting the active energy of light into a latent form that can be stored (in food) and used when needed. Photosynthesis provides us with most of the oxygen we need in order to breathe. We, in turn, exhale the carbon dioxide needed by plants. Plants are also crucial to human life because we rely on them as a source of food for ourselves and for the animals that we eat. There are two types of photosynthetic processes: oxygenic photosynthesis and anoxygenic photosynthesis. Oxygenic photosynthesis is the most common and is seen in plants, algae and cyanobacteria. On the other hand, anoxygenic photosynthesis uses electron donors other than water. The process typically occurs in bacteria such as purple bacteria and green sulfur bacteria. Plants make simple sugar by photosynthesis. It all happens in a fraction of a second. Water gets split into hydrogen and oxygen. The hydrogen is combined chemically with carbon dioxide to produce simple sugar. This dissolves easily in water and hence is transported to all parts of the plant. Oxygen is given off as a by-product. Thus, plants take in CO2 and gives out oxygen during photosynthesis reaction. Sun shines each day and does not charge us any money for the light it gives us. The light falling on Earth is very clean and will be there for a very long time as long as the Sun lasts. In addition to the current applications, Photosynthesis-based technology could also include using the concept of efficient "energy capture" to our artificial systems just as plants have been doing: thousands of chlorophyll a molecules (antenna) serving one center where the process occurs efficiently; the use of compounds, produced by plants, in triggering reactions that kill cancer cells; and the use of photosynthetic organisms in cleaning of aqueous surface environments (lakes, etc). An excellent example is in the use of cyanobacteria that literally eat up the nitrates from ground water and clear it for us. The opportunities that photosynthesis-based technology provides us are enormous. This process is extremely important for life on earth as it provides the oxygen that all other life depends on.

Content Coverage

Chapters One and *Two* introduce about photosynthesis process and chlorophyll, correspondingly. Chlorophyll is a complex organic molecule that enables plants, and some other organisms, to carry out photosynthesis — the conversion of carbon dioxide and water to glucose and oxygen using sunlight.

Chapters Three and *Four* present the role of chloroplast and light-dependent reactions, respectively. In photosynthesis, the light-dependent reactions take place on the thylakoid membranes. The inside of the thylakoid membrane is called the lumen, and outside the thylakoid membrane is the stroma, where the light-independent reactions take place.

Chapter Five gives an overview of photodissociation, photolysis, or photodecomposition. Photolysis is part of the light-dependent reactions of photosynthesis. The effectiveness of photons of different wavelengths depends on the absorption spectra of the photosynthetic pigments in the organism. Chlorophylls absorb light in the violet-blue and red parts of the spectrum, while accessory pigments capture other wavelengths as well.

Chapter Six focuses on carbon fixation or carbon assimilation, which is the conversion process of inorganic carbon (carbon dioxide) to organic compounds by living organisms. The most prominent example is photosynthesis, although chemosynthesis is another form of carbon fixation that can take place in the absence of sunlight.

Chapters Seven and *Eight* present the process of CAM photosynthesis and RuBisCO, separately. Crassulacean acid metabolism (CAM) is a photosynthetic CO2 fixation pathway that maximizes water use efficiency (WUE) many times relative to C3 species by using a temporal CO2 pump. RuBisCO, however, is an enzyme involved in the first major step of carbon fixation, a process by which atmospheric carbon dioxide is converted by plants and other photosynthetic organisms to energy-rich molecules such as glucose.

Chapters Nine and *Ten* introduces about the Emerson and Warberg Effect. The Emerson effect is the increase in the rate of photosynthesis after chloroplasts are exposed to light of wavelength 670 nm (red light) and 700 nm (far red light). However, the Warburg's effect is the decrease in the rate of photosynthesis by high oxygen concentrations.

CHAPTER 1

Photosynthesis: Introduction and Evolution

Chapter Contents

- **Introduction**
- **Overview**
- **Evolution**

Photosynthesis is a technique for converting sunlight into energy that has been utilized by certain organisms for around 3.4 billion years. The basic formula involves a conversion of carbon dioxide and water into glucose and oxygen, aided by light-dependent reactions. Photosynthesis emerged quickly after the emergence of life itself, thought to have occurred 3.5 billion years ago, after the Earth's crust cooled. The first photosynthetic organisms were the ancestors of modern-day cyanobacteria.

INTRODUCTION

Photosynthesis is a process used by plants and other organisms to convert light energy into chemical energy that can later be released to fuel the organisms' activities (energy transformation). This chemical energy is stored in carbohydrate molecules, such as sugars, which are synthesized from carbon dioxide and water – hence the name *photosynthesis*, from the Greek φῶς, phōs,, "light", and σύνθεσις, *synthesis*, "putting together". In most cases, oxygen is also released as a waste product. Most plants, most algae, and cyanobacteria perform photosynthesis; such organisms are called photoautotrophs. Photosynthesis is largely responsible for producing and maintaining the oxygen content of the Earth's atmosphere, and supplies all of the organic compounds and most of the energy necessary for life on Earth.

Although photosynthesis is performed differently by different species, the process always begins when energy from light is absorbed by proteins called reaction centres that contain green chlorophyll pigments. In plants, these proteins are held inside organelles called chloroplasts, which are most abundant in leaf cells, while in bacteria they are embedded in the plasma membrane. In these light-dependent reactions, some energy is used to strip electrons from suitable substances, such as water, producing oxygen gas. The hydrogen freed by the splitting of water is used in the creation of two further compounds that act as an immediate energy storage means: reduced nicotinamide adenine dinucleotide phosphate (NADPH) and adenosine triphosphate (ATP), the "energy currency" of cells.

In plants, algae and cyanobacteria, long-term energy storage in the form of sugars is produced by a subsequent sequence of light-independent reactions called the Calvin cycle; some bacteria use different mechanisms, such as the reverse Krebs cycle, to achieve the same end. In the Calvin cycle, atmospheric carbon dioxide is incorporated into already existing organic carbon compounds, such as ribulose bisphosphate (RuBP). Using the ATP and NADPH produced by the light-dependent reactions, the resulting compounds are then reduced and removed to form further carbohydrates, such as glucose.

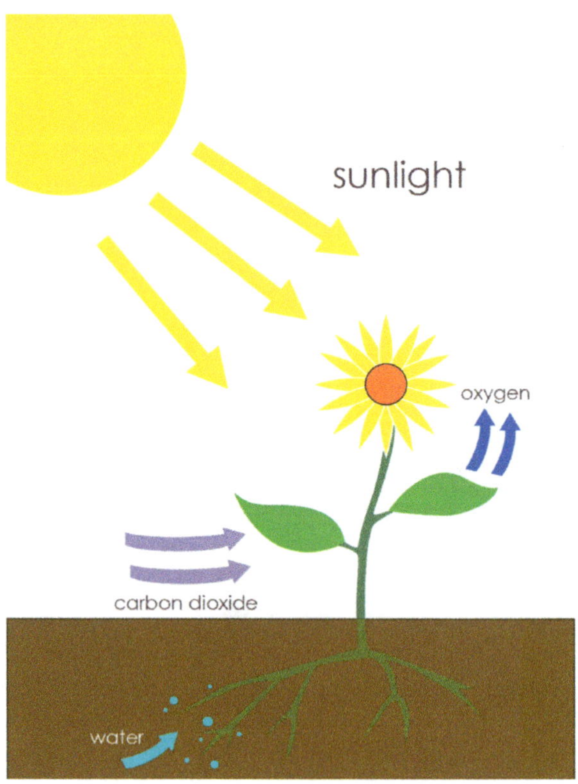

Schematic of photosynthesis in plants. The carbohydrates produced are stored in or used by the plant.

The first photosynthetic organisms probably evolved early in the evolutionary history of life and most likely used reducing agents such as hydrogen or hydrogen sulfide, rather than water, as sources of electrons. Cyanobacteria appeared later; the excess oxygen they produced contributed directly to the oxygenation of the Earth, which rendered the evolution of complex life possible. Today, the average rate of energy capture by photosynthesis globally is approximately 130 terawatts, which is about three times the current power consumption of human civilization. Photosynthetic organisms also convert around 100–115 thousand million metric tonnes of carbon into biomass per year.

$$\underset{\text{Carbon dioxide}}{6CO_2} + \underset{\text{Water}}{6H_2O} \xrightarrow{\text{Light}} \underset{\text{Sugar}}{C_6H_{12}O_6} + \underset{\text{Oxygen}}{6O_2}$$

Overall equation for the type of photosynthesis that occurs in plants

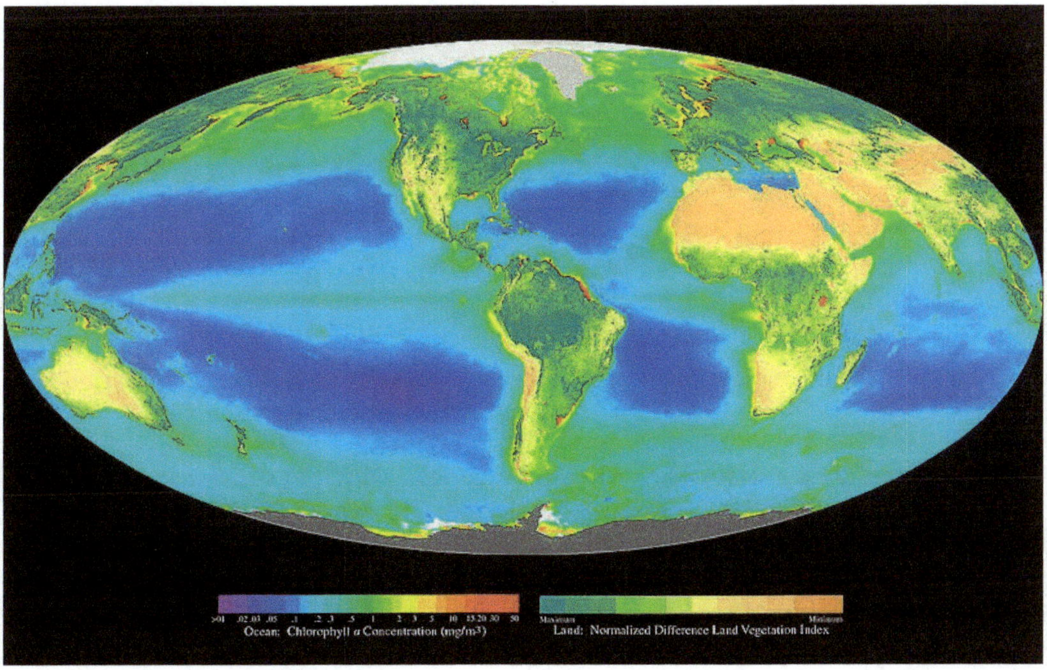

Composite image showing the global distribution of photosynthesis, including both oceanic phytoplankton and terrestrial vegetation. Dark red and blue-green indicate regions of high photosynthetic activity in the ocean and on land, respectively.

OVERVIEW

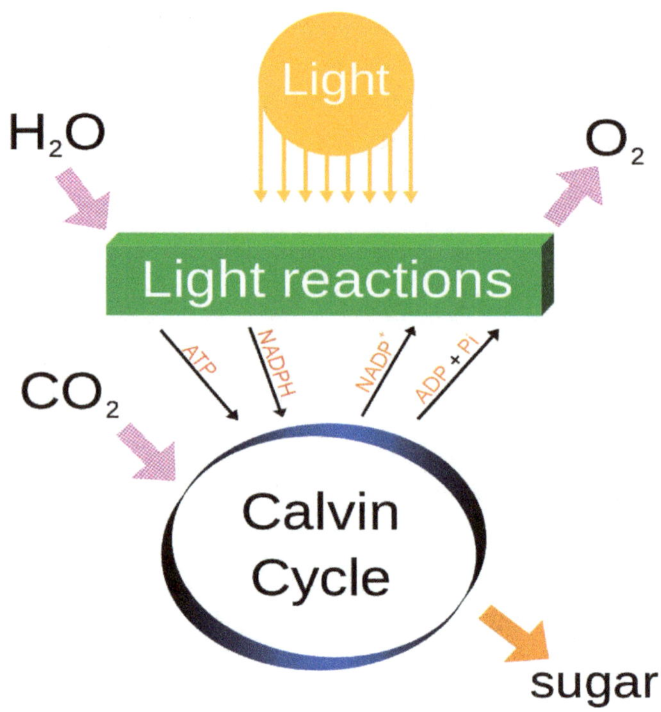

Photosynthesis changes sunlight into chemical energy, splits water to liberate O_2, and fixes CO_2 into sugar.

Photosynthetic organisms are photoautotrophs, which means that they are able to synthesize food directly from carbon dioxide and water using energy from light. However, not all organisms that use light as a source of energy carry out photosynthesis; photoheterotrophs use organic compounds, rather than carbon dioxide, as a source of carbon. In plants, algae, and cyanobacteria, photosynthesis releases oxygen. This is called *oxygenic photosynthesis* and is by far the most common type of photosynthesis used by living organisms. Although there are some differences between oxygenic photosynthesis in plants, algae, and cyanobacteria, the overall process is quite similar in these organisms. There are also many varieties of anoxygenic photosynthesis, used mostly by certain types of bacteria, which consume carbon dioxide but do not release oxygen.

Carbon dioxide is converted into sugars in a process called carbon fixation. Carbon fixation is an endothermic redox reaction, so photosynthesis needs to supply both a source of energy to drive this process, and the electrons needed to convert carbon dioxide into a carbohydrate via a reduction reaction. The addition of electrons to a chemical species is called reduction. In general outline and in effect, photosynthesis is the opposite of cellular respiration, in which glucose and other compounds are oxidized to produce carbon dioxide and water, and to release chemical energy (an exothermic reaction) to drive the organism›s metabolism. The two processes, reduction of carbon dioxide to carbohydrate and then later oxidation of the carbohydrate, are distinct: photosynthesis and cellular respiration take place through a different sequence of chemical reactions and in different cellular compartments.

The general equation for photosynthesis as first proposed by Cornelius van Niel is therefore:

$$CO_2 + 2H_2A + \text{photons} \rightarrow [CH_2O] + 2A + H_2O$$

carbon dioxide + electron donor + light energy → carbohydrate + oxidized electron donor + water

Since water is used as the electron donor in oxygenic photosynthesis, the equation for this process is:

$$CO_2 + 2H_2O + \text{photons} \rightarrow [CH_2O] + O_2 + H_2O$$

carbon dioxide + water + light energy → carbohydrate + oxygen + water

This equation emphasizes that water is both a reactant in the light-dependent reaction and a product of the light-independent reaction, but canceling *n* water molecules from each side gives the net equation:

$$CO_2 + H_2O + \text{photons} \rightarrow [CH_2O] + O_2$$

carbon dioxide + water + light energy → carbohydrate + oxygen

Other processes substitute other compounds (such as arsenite) for water in the electron-supply role; for example some microbes use sunlight to oxidize arsenite to arsenate: The equation for this reaction is:

$$CO_2 + (AsO3^{-3}) + \text{photons} \rightarrow (AsO3^{-4}) + CO$$

carbon dioxide + arsenite + light energy → arsenate + carbon monoxide (used to build other compounds in subsequent reactions)

Photosynthesis occurs in two stages. In the first stage, *light-dependent reactions* or *light reactions* capture the energy of light and use it to make the energy-storage molecules ATP and NADPH. During the second stage, the *light-independent reactions* use these products to capture and reduce carbon dioxide.

Most organisms that utilize oxygenic photosynthesis use visible light for the light-dependent reactions, although at least three use shortwave infrared or, more specifically, far-red radiation.

Some organisms employ even more radical variants of photosynthesis. Some archea use a simpler method that employs a pigment similar to those used for vision in animals. The bacteriorhodopsin changes its configuration in response to sunlight, acting as a proton pump. This produces a proton gradient more directly, which is then converted to chemical energy. The process does not involve carbon dioxide fixation and does not release oxygen, and seems to have evolved separately from the more common types of photosynthesis.

Photosynthetic Membranes and Organelles

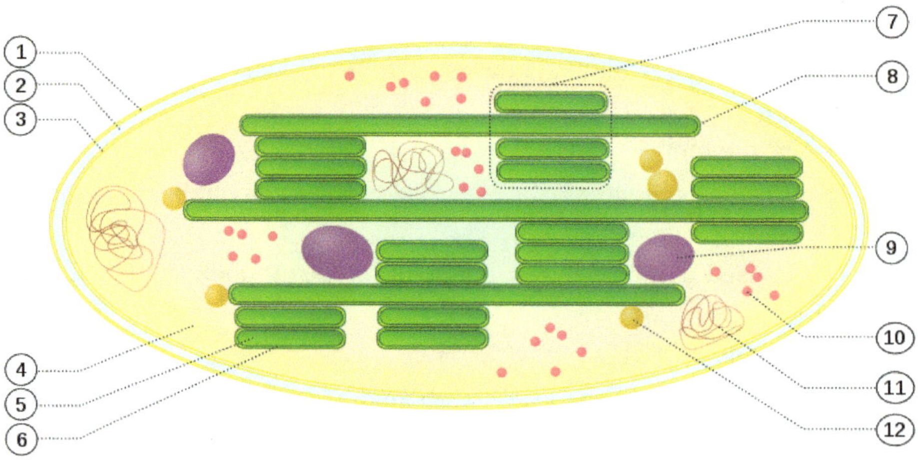

Chloroplast ultrastructure

1. outer membrane
2. intermembrane space
3. inner membrane (1+2+3: envelope)
4. stroma (aqueous fluid)
5. thylakoid lumen (inside of thylakoid)
6. thylakoid membrane
7. granum (stack of thylakoids)
8. thylakoid (lamella)
9. starch
10. ribosome
11. plastidial DNA
12. plastoglobule (drop of lipids)

In photosynthetic bacteria, the proteins that gather light for photosynthesis are embedded in cell membranes. In its simplest form, this involves the membrane surrounding the cell itself. However, the membrane may be tightly folded into cylindrical sheets called thylakoids, or bunched up into round vesicles called *intracytoplasmic membranes*. These structures can fill most of the interior of a cell, giving the membrane a very large surface area and therefore increasing the amount of light that the bacteria can absorb.

In plants and algae, photosynthesis takes place in organelles called chloroplasts. A typical plant cell contains about 10 to 100 chloroplasts. The chloroplast is enclosed by a membrane. This membrane is composed of a phospholipid inner membrane, a phospholipid outer membrane, and an intermembrane space. Enclosed by the membrane is an aqueous fluid called the stroma. Embedded within the stroma are stacks of thylakoids (grana), which are the site of photosynthesis. The thylakoids appear as flattened disks. The thylakoid itself is enclosed by the thylakoid membrane, and within the enclosed volume is a lumen or thylakoid space. Embedded in the thylakoid membrane are integral and peripheral membrane protein complexes of the photosynthetic system.

Plants absorb light primarily using the pigment chlorophyll. The green part of the light spectrum is not absorbed but is reflected which is the reason that most plants have a green color. Besides chlorophyll, plants also use pigments such as carotenes and xanthophylls. Algae also use chlorophyll, but various other pigments are present, such as phycocyanin, carotenes, and xanthophylls in green algae, phycoerythrin in red algae (rhodophytes) and fucoxanthin in brown algae and diatoms resulting in a wide variety of colors.

These pigments are embedded in plants and algae in complexes called antenna proteins. In such proteins, the pigments are arranged to work together. Such a combination of proteins is also called a light-harvesting complex.

Although all cells in the green parts of a plant have chloroplasts, the majority of those are found in specially adapted structures called leaves. Certain species adapted to conditions of strong sunlight and aridity, such as many Euphorbia and cactus species, have their main photosynthetic organs in their stems. The cells in the interior tissues of a leaf, called the mesophyll, can contain between 450,000 and 800,000 chloroplasts for every square millimeter of leaf. The surface of the leaf is coated with a water-resistant waxy cuticle that protects the leaf from excessive evaporation of water and decreases the absorption of ultraviolet or blue light to reduce heating. The transparent epidermis layer allows light to pass through to the palisade mesophyll cells where most of the photosynthesis takes place.

Light-Dependent Reactions

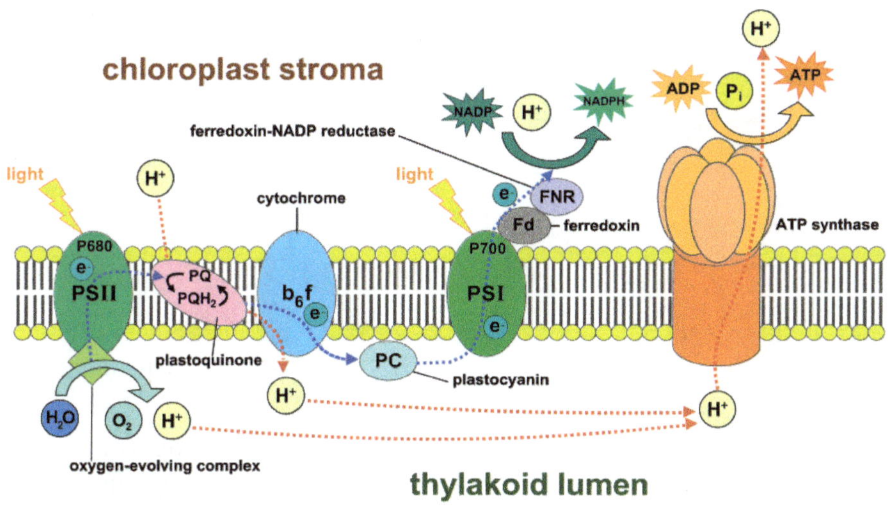

Light-dependent reactions of photosynthesis at the thylakoid membrane

In the light-dependent reactions, one molecule of the pigment chlorophyll absorbs one photon and loses one electron. This electron is passed to a modified form of chlorophyll called pheophytin, which passes the electron to a quinone molecule, starting the flow of electrons down an electron transport chain that leads to the ultimate reduction of NADP to NADPH. In addition, this creates a proton gradient (energy gradient) across the chloroplast membrane, which is used by ATP synthase in the synthesis of ATP. The chlorophyll molecule ultimately regains the electron it lost when a water molecule is split in a process called photolysis, which releases a dioxygen (O_2) molecule as a waste product.

The overall equation for the light-dependent reactions under the conditions of non-cyclic electron flow in green plants is:

$$2\ H_2O + 2\ NADP^+ + 3\ ADP + 3\ P_i + \text{light} \rightarrow 2\ NADPH + 2\ H^+ + 3\ ATP + O_2$$

Not all wavelengths of light can support photosynthesis. The photosynthetic action spectrum depends on the type of accessory pigments present. For example, in green plants, the action spectrum resembles the absorption spectrum for chlorophylls and carotenoids with peaks for violet-blue and red light. In red algae, the action spectrum is blue-green light, which allows these algae to use the blue end of the spectrum to grow in the deeper waters that filter out the longer wavelengths (red light) used by above ground green plants. The non-absorbed part of the light spectrum is what gives photosynthetic organisms their color (e.g., green plants, red algae, purple bacteria) and is the least effective for photosynthesis in the respective organisms.

Z scheme

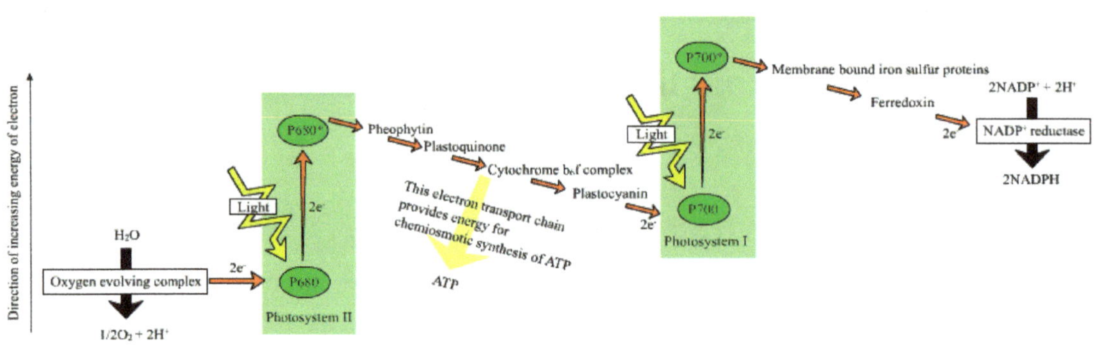

The "Z scheme"

In plants, light-dependent reactions occur in the thylakoid membranes of the chloroplasts where they drive the synthesis of ATP and NADPH. The light-dependent reactions are of two forms: cyclic and non-cyclic.

In the non-cyclic reaction, the photons are captured in the light-harvesting antenna complexes of photosystem II by chlorophyll and other accessory pigments (see diagram at right). The absorption of a photon by the antenna complex frees an electron by a process called photoinduced charge separation. The antenna system is at the core of the chlorophyll molecule of the photosystem II reaction center. That freed electron is transferred to the primary electron-acceptor molecule, pheophytin. As the electrons are shuttled through an electron transport chain (the so-called ***Z-scheme*** shown in the diagram), it initially functions to generate a chemiosmotic potential by pumping proton cations (H^+) across the membrane and into the thylakoid space. An ATP synthase enzyme uses that chemiosmotic potential to make ATP during photophosphorylation, whereas NADPH is a product of the terminal redox reaction in the *Z-scheme*. The electron enters a chlorophyll molecule in Photosystem I. There it is further excited by the light absorbed by that photosystem. The electron is then passed along a chain of electron acceptors to which it transfers some of its energy. The energy delivered to the electron acceptors is used to move hydrogen ions across the thylakoid membrane into the lumen. The electron is eventually used to reduce the co-enzyme NADP with a H^+ to NADPH (which has functions in the light-independent reaction); at that point, the path of that electron ends.

The cyclic reaction is similar to that of the non-cyclic, but differs in that it generates only ATP, and no reduced NADP (NADPH) is created. The cyclic reaction takes place only at photosystem I. Once the electron is displaced from the photosystem, the electron is passed down the electron acceptor molecules and returns to photosystem I, from where it was emitted, hence the name *cyclic reaction*.

Water Photolysis

The NADPH is the main reducing agent produced by chloroplasts, which then goes on to provide a source of energetic electrons in other cellular reactions. Its production leaves chlorophyll in photosystem I with a deficit of electrons (chlorophyll has been oxidized), which must be balanced by some other reducing agent that will supply the missing electron. The excited electrons lost from chlorophyll from photosystem I are supplied from the electron transport chain by plastocyanin. However, since photosystem II is the first step of the *Z-scheme*, an external source of electrons is required to reduce its oxidized chlorophyll *a* molecules. The source of electrons in green-plant and cyanobacterial photosynthesis is water. Two water molecules are oxidized by four successive charge-separation reactions by photosystem II to yield a molecule of diatomic oxygen and four hydrogen ions; the electrons yielded are transferred to a redox-active tyrosine residue that then reduces the oxidized chlorophyll *a* (called P680) that serves as the primary light-driven electron donor in the photosystem II reaction center. That photo receptor is in effect reset and is then able to repeat the absorption of another photon and the release of another photo-dissociated electron. The oxidation of water is catalyzed in photosystem II by a redox-active structure that contains four manganese ions and a calcium ion; this oxygen-evolving complex binds two water molecules and contains the four oxidizing equivalents that are used to drive the water-oxidizing reaction. Photosystem II is the only known biological enzyme that carries out this oxidation of water. The hydrogen ions released contribute to the transmembrane chemiosmotic potential that leads to ATP synthesis. Oxygen is a waste product of light-dependent reactions, but the majority of organisms on Earth use oxygen for cellular respiration, including photosynthetic organisms.

Light-Independent Reactions

Calvin Cycle

In the light-independent (or "dark") reactions, the enzyme RuBisCO captures CO_2 from the atmosphere and, in a process called the Calvin-Benson cycle, it uses the newly formed NADPH and releases three-carbon sugars, which are later combined to form sucrose and starch. The overall equation for the light-independent reactions in green plants is

$$3\ CO_2 + 9\ ATP + 6\ NADPH + 6\ H^+ \rightarrow C_3H_6O_3\text{-phosphate} + 9\ ADP + 8\ P_i + 6\ NADP^+ + 3\ H_2O$$

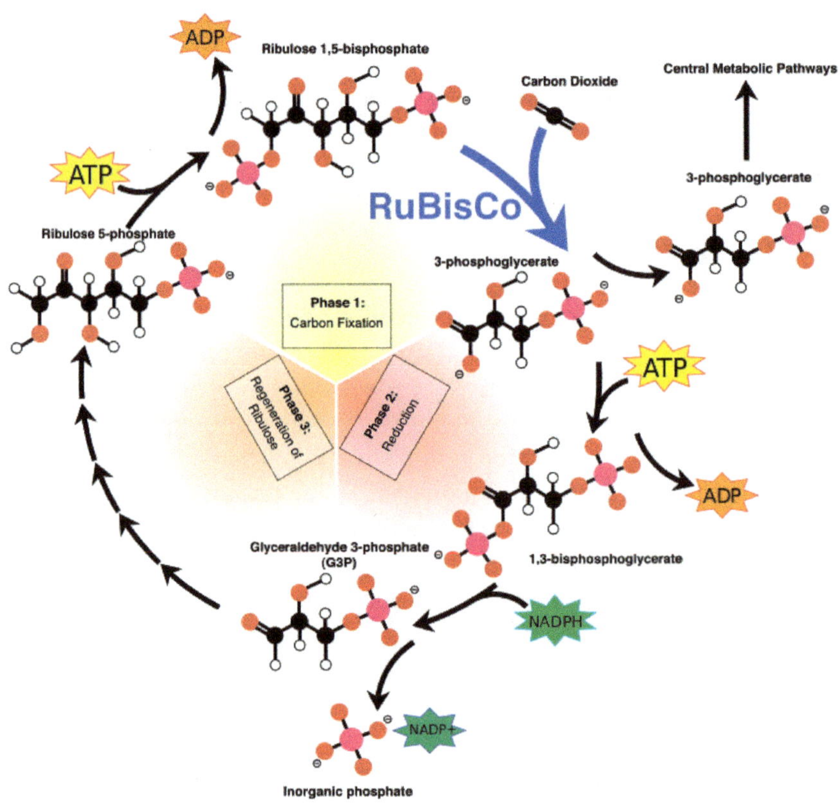

Overview of the Calvin cycle and carbon fixation

Carbon fixation produces the intermediate three-carbon sugar product, which is then converted to the final carbohydrate products. The simple carbon sugars produced by photosynthesis are then used in the forming of other organic compounds, such as the building material cellulose, the precursors for lipid and amino acid biosynthesis, or as a fuel in cellular respiration. The latter occurs not only in plants but also in animals when the energy from plants is passed through a food chain.

The fixation or reduction of carbon dioxide is a process in which carbon dioxide combines with a five-carbon sugar, ribulose 1,5-bisphosphate, to yield two molecules of a three-carbon compound, glycerate 3-phosphate, also known as 3-phosphoglycerate. Glycerate 3-phosphate, in the presence of ATP and NADPH produced during the light-dependent stages, is reduced to glyceraldehyde 3-phosphate. This product is also referred to as 3-phosphoglyceraldehyde (PGAL) or, more generically, as triose phosphate. Most (5 out of 6 molecules) of the glyceraldehyde 3-phosphate produced is used to regenerate ribulose 1,5-bisphosphate so the process can continue. The triose phosphates not thus "recycled" often condense to form hexose phosphates, which ultimately yield sucrose, starch and cellulose. The sugars produced during carbon metabolism yield carbon skeletons that can be used for other metabolic reactions like the production of amino acids and lipids.

Carbon Concentrating Mechanisms

On Land

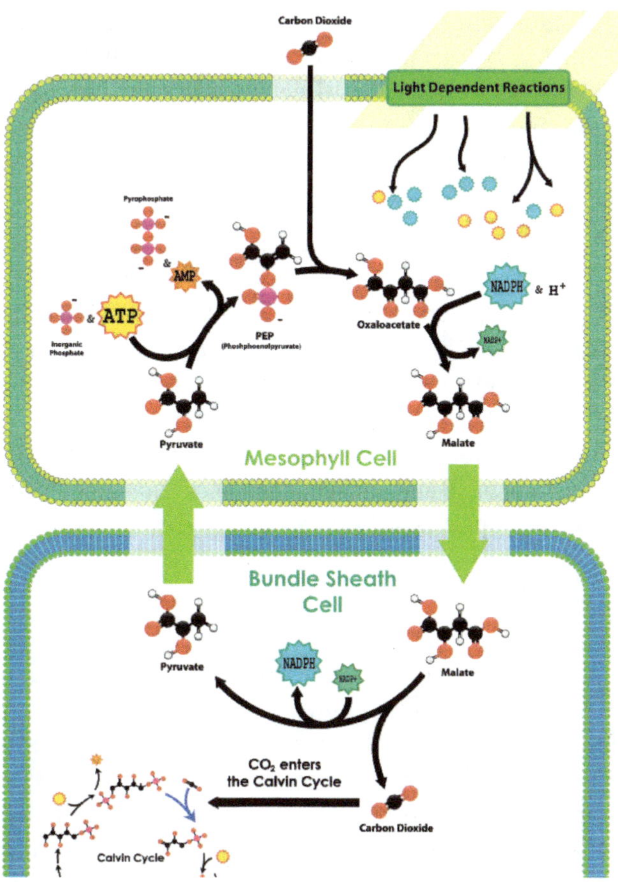

Overview of C4 carbon fixation

In hot and dry conditions, plants close their stomata to prevent water loss. Under these conditions, CO_2 will decrease and oxygen gas, produced by the light reactions of photosynthesis, will increase, causing an increase of photorespiration by the oxygenase activity of ribulose-1,5-bisphosphate carboxylase/oxygenase and decrease in carbon fixation. Some plants have evolved mechanisms to increase the CO_2 concentration in the leaves under these conditions.

Plants that use the C_4 carbon fixation process chemically fix carbon dioxide in the cells of the mesophyll by adding it to the three-carbon molecule phosphoenolpyruvate (PEP), a reaction catalyzed by an enzyme called PEP carboxylase, creating the four-carbon organic acid oxaloacetic acid. Oxaloacetic acid or malate synthesized by this process is then translocated to specialized bundle sheath cells where the enzyme RuBisCO and other Calvin cycle enzymes are located, and where CO_2 released by decarboxylation of the four-carbon acids is then fixed by RuBisCO activity to the three-carbon 3-phosphoglyceric acids. The physical separation of RuBisCO from the oxygen-generating light reactions reduces photorespiration and increases CO_2 fixation and, thus, the photosynthetic capacity of the leaf. C_4 plants can produce more sugar than C_3 plants in conditions of high light and temperature. Many important crop plants are C_4 plants, including maize, sorghum, sugarcane, and millet. Plants that do not use PEP-carboxylase in carbon fixation are called C_3 plants because the primary carboxylation reaction, catalyzed by RuBisCO, produces the three-carbon 3-phosphoglyceric acids directly in the Calvin-Benson cycle. Over 90% of plants use C_3 carbon fixation, compared to 3% that use C_4carbon fixation; however, the evolution of C_4 in over 60 plant lineages makes it a striking example of convergent evolution.

Xerophytes, such as cacti and most succulents, also use PEP carboxylase to capture carbon dioxide in a process called Crassulacean acid metabolism (CAM). In contrast to C_4 metabolism, which *spatially* separates the CO_2 fixation to PEP from the Calvin cycle, CAM *temporally* separates these two processes. CAM plants have a different leaf anatomy from C_3 plants, and fix the CO_2 at night, when their stomata are open. CAM plants store the CO_2 mostly in the form of malic acid via carboxylation of phosphoenolpyruvate to oxaloacetate, which is then reduced to malate. Decarboxylation of malate during the day releases CO_2 inside the leaves, thus allowing carbon fixation to 3-phosphoglycerate by RuBisCO. Sixteen thousand species of plants use CAM.

In Water

Cyanobacteria possess carboxysomes, which increase the concentration of CO_2 around RuBisCO to increase the rate of photosynthesis. An enzyme, carbonic anhydrase, located within the carboxysome releases CO_2 from the dissolved hydrocarbonate ions (HCO^{-3}). Before the CO_2 diffuses out it is quickly sponged up by RuBisCO, which is concentrated within the carboxysomes. HCO^{-3} ions are made from CO_2 outside the cell by another carbonic anhydrase and are actively pumped into the cell by a membrane protein. They cannot cross the membrane as they are charged, and within the cytosol they turn back into CO_2 very slowly without the help of carbonic anhydrase. This causes the $HCO^{-}3$ ions to accumulate within the cell from where they diffuse into the carboxysomes. Pyrenoids in algae and hornworts also act to concentrate CO_2 around rubisco.

Order and Kinetics

The overall process of photosynthesis takes place in four stages:

Stage	Description	Time scale
1	Energy transfer in antenna chlorophyll (thylakoid membranes)	femtosecond to picosecond
2	Transfer of electrons in photochemical reactions (thylakoid membranes)	picosecond to nanosecond
3	Electron transport chain and ATP synthesis (thylakoid membranes)	microsecond to millisecond
4	Carbon fixation and export of stable products	millisecond to second

Efficiency

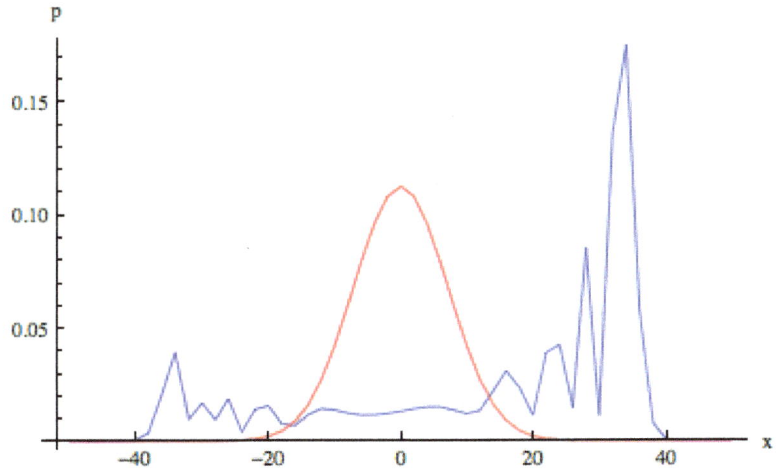

Probability distribution resulting from one-dimensional discrete time random walks. The quantum walk created using the Hadamard coin is plotted (blue) vs a classical walk (red) after 50 time steps.

Plants usually convert light into chemical energy with a photosynthetic efficiency of 3–6%. Absorbed light that is unconverted is dissipated primarily as heat, with a small fraction (1–2%) re-emitted as chlorophyll fluorescence at longer (redder) wavelengths. A fact that allows measurement of the light reaction of photosynthesis by using chlorophyll fluorometers.

Actual plants' photosynthetic efficiency varies with the frequency of the light being converted, light intensity, temperature and proportion of carbon dioxide in the atmosphere, and can vary from 0.1% to 8%. By comparison, solar panels convert light into electric energy at an efficiency of approximately 6–20% for mass-produced panels, and above 40% in laboratory devices.

The efficiency of both light and dark reactions can be measured but the relationship between the two can be complex. For example, the ATP and NADPH energy molecules, created by the light reaction, can be used for carbon fixation or for photorespiration in C_3 plants. Electrons may also flow to other electron sinks. For this reason, it is not uncommon for authors to differentiate between work done under non-photorespiratory conditions and under photorespiratory conditions.

Chlorophyll fluorescence of photosystem II can measure the light reaction, and Infrared gas analyzers can measure the dark reaction. It is also possible to investigate both at the same time using an integrated chlorophyll fluorometer and gas exchange system, or by using two separate systems together. Infrared gas analyzers and some moisture sensors are sensitive enough to measure the photosynthetic assimilation of CO_2, and of ΔH_2O using reliable methods CO_2 is commonly measured in μmols/m²/s⁻¹, parts per million or volume per million and H_20 is commonly measured in mmol/m²/s⁻¹ or in mbars. By measuring CO_2assimilation, ΔH_2O, leaf temperature, barometric pressure, leaf area, and photosynthetically active radiation or PAR, it becomes possible to estimate, "A" or carbon assimilation, "E" or transpiration, "gs" or stomatal conductance, and Ci or intracellular CO_2. However, it is more common to used chlorophyll fluorescence for plant stress measurement, where appropriate, because the most commonly used measuring parameters FV/FM and Y(II) or F/FM' can be made in a few seconds, allowing the measurement of larger plant populations.

Gas exchange systems that offer control of CO_2 levels, above and below ambient, allow the common practice of measurement of A/Ci curves, at different CO_2levels, to characterize a plant's photosynthetic response.

Integrated chlorophyll fluorometer – gas exchange systems allow a more precise measure of photosynthetic response and mechanisms. While standard gas exchange photosynthesis systems can measure Ci, or substomatal CO_2 levels, the addition of integrated chlorophyll fluorescence measurements allows a more precise measurement of C_C to replace Ci. The estimation of CO_2 at the site of carboxylation in the chloroplast, or C_C, becomes possible with the measurement of mesophyll conductance or g_m using an integrated system.

Photosynthesis measurement systems are not designed to directly measure the amount of light absorbed by the leaf. But analysis of chlorophyll-fluorescence, P700- and P515-absorbance and gas exchange measurements reveal detailed information about e.g. the photosystems, quantum efficiency and the CO_2assimilation rates. With some instruments even wavelength-dependency of the photosynthetic efficiency can be analyzed.

A phenomenon known as quantum walk increases the efficiency of the energy transport of light significantly. In the photosynthetic cell of an algae, bacterium, or plant, there are light-sensitive molecules called chromophores arranged in an antenna-shaped structure named a photocomplex. When a photon is absorbed by a chromophore, it is converted into a quasiparticle referred to as an exciton, which jumps from chromophore to chromophore towards the reaction center of the photocomplex, a collection of molecules that traps its energy in a chemical form that makes it accessible for the cell's metabolism. The exciton's wave properties enable it to cover a wider area and try out several possible paths simultaneously, allowing it to instantaneously "choose" the most efficient route, where it will have the highest probability of arriving at its destination in the minimum possible time. Because that quantum walking takes place at temperatures far higher than quantum phenomena usually occur, it is only possible over very short distances, due to obstacles in the form of destructive interference that come into play. These obstacles cause the particle to lose its wave properties for an instant before it regains them once again after it is freed from its locked position through a classic "hop". The movement of the electron towards the photo center is therefore covered in a series of conventional hops and quantum walks.

EVOLUTION

Early photosynthetic systems, such as those in green and purple sulfur and green and purple nonsulfur bacteria, are thought to have been anoxygenic, and used various other molecules as electron donors rather than water. Green and purple sulfur bacteria are thought to have used hydrogen and sulfur as electron donors. Green nonsulfur bacteria used various amino and other organic acids as an electron donor. Purple nonsulfur bacteria used a variety of nonspecific organic molecules. The use of these molecules is consistent with the geological evidence that Earth's early atmosphere was highly reducing at that time.

Fossils of what are thought to be filamentous photosynthetic organisms have been dated at 3.4 billion years old.

The main source of oxygen in the Earth's atmosphere derives from oxygenic photosynthesis, and its first appearance is sometimes referred to as the oxygen catastrophe. Geological evidence suggests that oxygenic photosynthesis, such as that in cyanobacteria, became important during the Paleoproterozoic era around 2 billion years ago. Modern photosynthesis in plants and most photosynthetic prokaryotes is oxygenic. Oxygenic photosynthesis uses water as an electron donor, which is oxidized to molecular oxygen (O_2) in the photosynthetic reaction center.

Symbiosis and the Origin of Chloroplasts

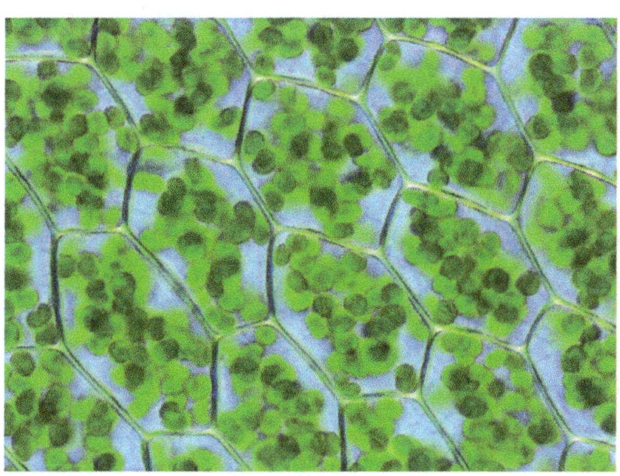

Plant cells with visible chloroplasts (from a moss, *Plagiomnium affine*)

Several groups of animals have formed symbiotic relationships with photosynthetic algae. These are most common in corals, sponges and sea anemones. It is presumed that this is due to the particularly simple body plans and large surface areas of these animals compared to their volumes. In addition, a few marine mollusks *Elysia viridis* and *Elysia chlorotica* also maintain a symbiotic relationship with chloroplasts they capture from the algae in their diet and then store in their bodies. This allows the mollusks to survive solely by photosynthesis for several months at a time. Some of the genes from the plant cell nucleus have even been transferred to the slugs, so that the chloroplasts can be supplied with proteins that they need to survive.

An even closer form of symbiosis may explain the origin of chloroplasts. Chloroplasts have many similarities with photosynthetic bacteria, including a circular chromosome, prokaryotic-type ribosome, and similar proteins in the photosynthetic reaction center. The endosymbiotic theory suggests that photosynthetic bacteria were acquired (by endocytosis) by early eukaryotic cells to form the first plant cells. Therefore, chloroplasts may be photosynthetic bacteria that adapted to life inside plant cells. Like mitochondria, chloroplasts possess their own DNA, separate from the nuclear DNA of their plant host cells and the genes in this chloroplast DNA resemble those found in cyanobacteria. DNA in chloroplasts codes for redox proteins such as those found in the photosynthetic reaction centers. The CoRR Hypothesis proposes that this **Co**-location is required for **R**edox **R**egulation.

Cyanobacteria and the Evolution of Photosynthesis

The biochemical capacity to use water as the source for electrons in photosynthesis evolved once, in a common ancestor of extant cyanobacteria. The geological record indicates that this transforming event took place early in Earth's history, at least 2450–2320 million years ago (Ma), and, it is speculated, much earlier.Because the Earth's atmosphere contained almost no oxygen during the estimated development of photosynthesis, it is believed that the first photosynthetic cyanobacteria did not generate oxygen. Available evidence from geobiological studies of Archean (>2500 Ma) sedimentary rocks indicates that life existed 3500 Ma, but the question of when oxygenic photosynthesis evolved is still unanswered. A clear paleontological window on cyanobacterial evolution opened about 2000 Ma, revealing an already-diverse biota of blue-green algae. Cyanobacteria remained the principal primary producers of oxygen throughout the Proterozoic Eon (2500–543 Ma), in part because the redox structure of the oceans favored photoautotrophs capable of nitrogen fixation. Green algae joined blue-green algae as the major primary producers of oxygen on continental shelves near the end of the Proterozoic, but it was only with the Mesozoic (251–65 Ma) radiations of dinoflagellates, coccolithophorids, and diatoms did the primary production of oxygen in marine shelf waters take modern form. Cyanobacteria remain critical to marine ecosystems as primary producers of oxygen in oceanic gyres, as agents of biological nitrogen fixation, and, in modified form, as the plastids of marine algae.

Discovery

Although some of the steps in photosynthesis are still not completely understood, the overall photosynthetic equation has been known since the 19th century.

Jan van Helmont began the research of the process in the mid-17th century when he carefully measured the mass of the soil used by a plant and the mass of the plant as it grew. After noticing that the soil mass changed very little, he hypothesized that the mass of the growing plant must come from the water, the only substance he added to the potted plant. His hypothesis was partially accurate — much of the gained mass also comes from carbon dioxide as well as water. However, this was a signaling point to the idea that the bulk of a plant's biomass comes from the inputs of photosynthesis, not the soil itself.

Joseph Priestley, a chemist and minister, discovered that, when he isolated a volume of air under an inverted jar, and burned a candle in it, the candle would burn out very quickly, much before it ran out of wax. He further discovered that a mouse could similarly "injure" air. He then showed that the air that had been "injured" by the candle and the mouse could be restored by a plant.

In 1778, Jan Ingenhousz, repeated Priestley's experiments. He discovered that it was the influence of sunlight on the plant that could cause it to revive a mouse in a matter of hours.

In 1796, Jean Senebier, a Swiss pastor, botanist, and naturalist, demonstrated that green plants consume carbon dioxide and release oxygen under the influence of light. Soon afterward, Nicolas-Théodore de Saussure showed that the increase in mass of the plant as it grows could not be due only to uptake of CO_2 but also to the incorporation of water. Thus, the basic reaction by which photosynthesis is used to produce food (such as glucose) was outlined.

Cornelis Van Niel made key discoveries explaining the chemistry of photosynthesis. By studying purple sulfur bacteria and green bacteria he was the first to demonstrate that photosynthesis is a light-dependent redox reaction, in which hydrogen reduces carbon dioxide.

Robert Emerson discovered two light reactions by testing plant productivity using different wavelengths of light. With the red alone, the light reactions were suppressed. When blue and red were combined, the output was much more substantial. Thus, there were two photosystems, one absorbing up to 600 nm wavelengths, the other up to 700 nm. The former is known as PSII, the latter is PSI. PSI contains only chlorophyll a , PSII contains primarily chlorophyll a with most of the available chlorophyll b , among other pigment. These include phycobilins, which are the red and blue pigments of red and blue algae respectively, and fucoxanthol for brown algae and diatoms. The process is most productive when the absorption of quanta are equal in both the PSII and PSI, assuring that input energy from the antenna complex is divided between the PSI and PSII system, which in turn powers the photochemistry.

Melvin Calvin works in his photosynthesis laboratory.

Robert Hill thought that a complex of reactions consisting of an intermediate to cytochrome b_6 (now a plastoquinone), another is from cytochrome f to a step in the carbohydrate-generating mechanisms. These are linked by plastoquinone, which does require energy to reduce cytochrome f for it is a sufficient reductant. Further experiments to prove that the oxygen developed during the photosynthesis of green plants came from water, were performed by Hill in 1937 and 1939. He showed that isolated chloroplasts give off oxygen in the presence of unnatural reducing agents like iron oxalate, ferricyanide or benzoquinone after exposure to light. The Hill reaction is as follows:

$$2\ H_2O + 2\ A + (\text{light, chloroplasts}) \rightarrow 2\ AH_2 + O_2$$

where A is the electron acceptor. Therefore, in light, the electron acceptor is reduced and oxygen is evolved.

Samuel Ruben and Martin Kamen used radioactive isotopes to determine that the oxygen liberated in photosynthesis came from the water.

Melvin Calvin and Andrew Benson, along with James Bassham, elucidated the path of carbon assimilation (the photosynthetic carbon reduction cycle) in plants. The carbon reduction cycle is known as the Calvin cycle, which ignores the contribution of Bassham and Benson. Many scientists refer to the cycle as the Calvin-Benson Cycle, Benson-Calvin, and some even call it the Calvin-Benson-Bassham (or CBB) Cycle.

Nobel Prize-winning scientist Rudolph A. Marcus was able to discover the function and significance of the electron transport chain.

Otto Heinrich Warburg and Dean Burk discovered the I-quantum photosynthesis reaction that splits the CO_2, activated by the respiration.

Louis N.M. Duysens and Jan Amesz discovered that chlorophyll a will absorb one light, oxidize cytochrome f, chlorophyll a (and other pigments) will absorb another light, but will reduce this same oxidized cytochrome, stating the two light reactions are in series.

Development of the Concept

In 1893, Charles Reid Barnes proposed two terms, photosyntax and photosynthesis, for the biological process of synthesis of complex carbon compounds out of carbonic acid, in the presence of chlorophyll, under the influence of light. Over time, the term photosynthesis came into common usage as the term of choice. Later discovery of anoxygenic photosynthetic bacteria and photophosphorylation necessitated redefinition of the term.

C3 : C4 Photosynthesis Research

After WWII at late 1940 at the University of California, Berkeley, the details of photosynthetic carbon metabolism were sorted out by the chemists Melvin Calvin, Andrew Benson, James Bassham and a score of students and researchers utilizing the carbon-14 isotope and paper chromatography techniques. The pathway of CO2 fixation by the algae *Chlorella* in a fraction of a second in light resulted in a 3 carbon molecule called phosphoglyceric acid (PGA). For that original and ground-breaking work, a Nobel Prize in Chemistry was awarded to Melvin Calvin 1961. In parallel, plant physiologists studied leaf gas exchanges using the new method of infrared gas analysis and a leaf chamber where the net photosynthetic rates ranged from 10 to 13 u mole CO2/square metere.sec., with the conclusion that all terrestrial plants having the same photosynthetic capacities that were light saturated at less than 50% of sunlight. These rates were determined in potted plants grown indoors under low light intensity.

Later in 1958-1963 at Cornell University, field grown maize was reported to have much greater leaf photosynthetic rates of 40 u mol CO2/square meter.sec and was not saturated at near full sunlight. This higher rate in maize was almost double those observed in other species such as wheat and soybean, indicating that large differences in photosynthesis exist among higher plants. At the University of Arizona, detailed gas exchange research on more than 15 species of monocot and dicot uncovered for the first time that differences in leaf anatomy are crucial factors in differentiating photosynthetic capacities among species. In tropical grasses, including maize, sorghum, sugarcane, Bermuda grass and in the dicot amaranthus, leaf photosynthetic rates were around 38–40 u mol CO2/square meter.sec., and the leaves have two types of green cells, i. e. outer layer of mesophyll cells surrounding a tightly packed cholorophyllous vascular bundle sheath cells. This type of anatomy was termed Kranz anatomy in the 19th century by the botanist Gottlieb Haberlandt while studying leaf anatomy of sugarcane. Plant species with the greatest photosynthetic rates and Kranz anatomy showed no apparent photorespiration, very low

CO_2 compensation point, high optimum temperature, high stomatal resistances and lower mesophyll resistances for gas diffusion and rates never saturated at full sun light. The research at Arizona was designated Citation Classic by the ISI 1986. These species was later termed C4 plants as the first stable compound of CO_2 fixation in light has 4 carbon as malate and aspartate. Other species that lack Kranz anatomy were termed C3 type such as cotton and sunflower, as the first stable carbon compound is the 3-carbon PGA acid. At 1000 ppm CO_2 in measuring air, both the C3 and C4 plants had similar leaf photosynthetic rates around 60 u mole CO_2/square meter.sec. indicating the suppression of phototorespiration in C3 plants.

Factors

The leaf is the primary site of photosynthesis in plants.

There are three main factors affecting photosynthesis and several corollary factors. The three main are:

- Light irradiance and wavelength
- Carbon dioxide concentration
- Temperature.

Light Intensity (Irradiance), Wavelength and Temperature

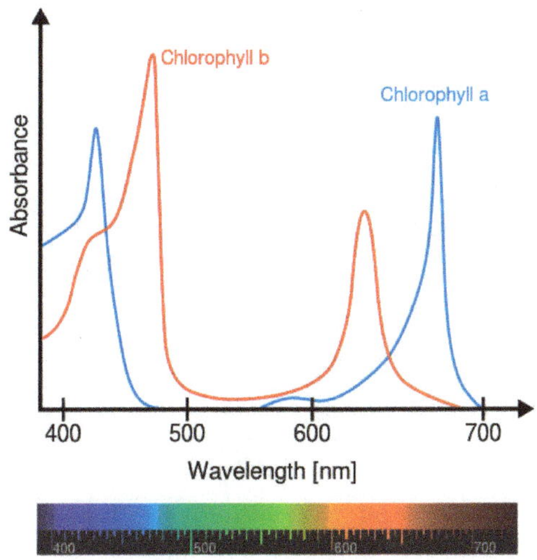

Absorbance spectra of free chlorophyll *a* (green) and *b* (red) in a solvent. The action spectra of chlorophyll molecules are slightly modified *in vivo* depending on specific pigment-protein interactions.

The process of photosynthesis provides the main input of free energy into the biosphere, and is one of four main ways in which radiation is important for plant life.

The radiation climate within plant communities is extremely variable, with both time and space.

In the early 20th century, Frederick Blackman and Gabrielle Matthaei investigated the effects of light intensity (irradiance) and temperature on the rate of carbon assimilation.

- At constant temperature, the rate of carbon assimilation varies with irradiance, increasing as the irradiance increases, but reaching a plateau at higher irradiance.
- At low irradiance, increasing the temperature has little influence on the rate of carbon assimilation. At constant high irradiance, the rate of carbon assimilation increases as the temperature is increased.

These two experiments illustrate several important points: First, it is known that, in general, photochemical reactions are not affected by temperature. However, these experiments clearly show that temperature affects the rate of carbon assimilation, so there must be two sets of reactions in the full process of carbon assimilation. These are, of course, the light-dependent ‹photochemical› temperature-independent stage, and the light-independent, temperature-dependent stage. Second, Blackman›s experiments illustrate the concept of limiting factors. Another limiting factor is the wavelength of light. Cyanobacteria, which reside several meters underwater, cannot receive the correct wavelengths required to cause photoinduced charge separation in conventional photosynthetic pigments. To combat this problem, a series of proteins with different pigments surround the reaction center. This unit is called a phycobilisome.

Carbon Dioxide Levels and Photorespiration

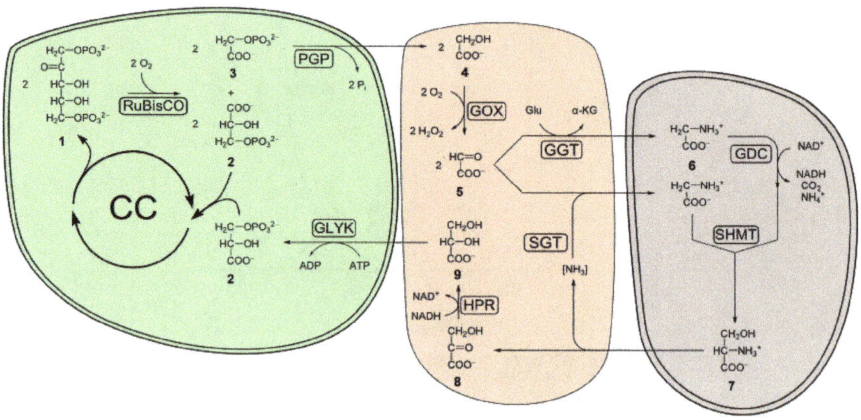

Photorespiration

As carbon dioxide concentrations rise, the rate at which sugars are made by the light-independent reactions increases until limited by other factors. RuBisCO, the enzyme that captures carbon dioxide in the light-independent reactions, has a binding affinity for both carbon dioxide and oxygen. When the concentration of carbon dioxide is high, RuBisCO will fix carbon dioxide. However, if the carbon dioxide concentration is low, RuBisCO will bind oxygen instead of carbon dioxide. This process, called photorespiration, uses energy, but does not produce sugars.

RuBisCO oxygenase activity is disadvantageous to plants for several reasons:

- One product of oxygenase activity is phosphoglycolate (2 carbon) instead of 3-phosphoglycerate (3 carbon). Phosphoglycolate cannot be metabolized by the Calvin-Benson cycle and represents carbon lost from the cycle. A high oxygenase activity, therefore, drains the sugars that are required to recycle ribulose 5-bisphosphate and for the continuation of the Calvin-Benson cycle.
- Phosphoglycolate is quickly metabolized to glycolate that is toxic to a plant at a high concentration; it inhibits photosynthesis.
- Salvaging glycolate is an energetically expensive process that uses the glycolate pathway, and only 75% of the carbon is returned to the Calvin-Benson cycle as 3-phosphoglycerate. The reactions also produce ammonia (NH_3), which is able to diffuse out of the plant, leading to a loss of nitrogen.

A highly simplified summary is:

2 glycolate + ATP → 3-phosphoglycerate + carbon dioxide + ADP + NH_3

The salvaging pathway for the products of RuBisCO oxygenase activity is more commonly known as photorespiration, since it is characterized by light-dependent oxygen consumption and the release of carbon dioxide.

REFERENCES

1. "Photosynthesis". Online Etymology Dictionary.
2. Liddell, Henry George; Scott, Robert; A Greek–English Lexicon at the Perseus Project
3. σύνθεσις. Liddell, Henry George; Scott, Robert; A Greek–English Lexicon at the Perseus Project
4. Bryant DA, Frigaard NU (Nov 2006). Prokaryotic photosynthesis and phototrophy illuminated . Trends in Microbiology. 14 (11): 488–96. doi:10.1016/j.tim.2006.09.001. PMID 16997562.
5. Reece J, Urry L, Cain M, Wasserman S, Minorsky P, Jackson R. Biology (International ed.). Upper Saddle River, NJ: Pearson Education. pp. 235, 244. ISBN 0-321-73975-2. This initial incorporation of carbon into organic compounds is known as carbon fixation.
6. Olson JM (May 2006). Photosynthesis in the Archean era . Photosynthesis Research. 88 (2): 109–17. doi:10.1007/s11120-006-9040-5. PMID 16453059.
7. Buick R (Aug 2008). When did oxygenic photosynthesis evolve? . Philosophical Transactions of the Royal Society of London, Series B. 363 (1504): 2731–43. doi:10.1098/rstb.2008.0041. PMC 2606769 . PMID 18468984.
8. Nealson KH, Conrad PG (Dec 1999). Life: past, present and future . Philosophical Transactions of the Royal Society of London, Series B. 354 (1392): 1923–39. doi:10.1098/rstb.1999.0532. PMC 1692713 . PMID 10670014.
9. Whitmarsh J, Govindjee (1999). The photosynthetic process . In Singhal GS, Renger G, Sopory SK, Irrgang KD, Govindjee. Concepts in photobiology: photosynthesis and photomorphogenesis. Boston: Kluwer Academic Publishers. pp. 11–51. ISBN 0-7923-5519-9. 100×1015 grams of carbon/year fixed by photosynthetic organisms, which is equivalent to 4×1018 kJ/yr = 4×1021 J/yr of free energy stored as reduced carbon.
10. Steger U, Achterberg W, Blok K, Bode H, Frenz W, Gather C, Hanekamp G, Imboden D, Jahnke M, Kost M, Kurz R, Nutzinger HG, Ziesemer T (2005). Sustainable development and innovation in the energy sector. Berlin: Springer. p. 32. ISBN 3-540-23103-X. The average global rate of photosynthesis is 130 TW.
11. World Consumption of Primary Energy by Energy Type and Selected Country Groups, 1980–2004 (XLS). Energy Information Administration. July 31, 2006. Retrieved 2007-01-20.
12. Field CB, Behrenfeld MJ, Randerson JT, Falkowski P (Jul 1998). Primary production of the biosphere: integrating terrestrial and oceanic components . Science. 281 (5374): 237–40. Bibcode:1998Sci...281..237F. doi:10.1126/science.281.5374.237. PMID 9657713.

13. Photosynthesis . McGraw-Hill Encyclopedia of Science & Technology. 13. New York: McGraw-Hill. 2007. ISBN 0-07-144143-3.
14. Whitmarsh J, Govindjee (1999). Chapter 2: The Basic Photosynthetic Process . In Singhal GS, Renger G, Sopory SK, Irrgang KD, Govindjee. Concepts in Photobiology: Photosynthesis and Photomorphogenesis. Boston: Kluwer Academic Publishers. p. 13. ISBN 978-0-7923-5519-9.
15. Anaerobic Photosynthesis, Chemical & Engineering News, 86, 33, August 18, 2008, p. 36
16. Kulp TR, Hoeft SE, Asao M, Madigan MT, Hollibaugh JT, Fisher JC, Stolz JF, Culbertson CW, Miller LG, Oremland RS (Aug 2008). Arsenic(III) fuels anoxygenic photosynthesis in hot spring biofilms from Mono Lake, California . Science. 321 (5891): 967–70. Bibcode:2008Sci...321..967K. doi:10.1126/science.1160799. PMID 18703741.
17. Scientists discover unique microbe in California›s largest lake . Retrieved 2009-07-20.
18. Plants: Diversity and Evolution, page 14, Martin Ingrouille, Bill Eddie
19. Evolution of Photosynthesis
20. Tavano CL, Donohue TJ (Dec 2006). Development of the bacterial photosynthetic apparatus . Current Opinion in Microbiology. 9 (6): 625–31. doi:10.1016/j.mib.2006.10.005. PMC 2765710 . PMID 17055774.
21. Mullineaux CW (1999). The thylakoid membranes of cyanobacteria: structure, dynamics and function . Australian Journal of Plant Physiology. 26 (7): 671–677. doi:10.1071/PP99027.
22. Sener MK, Olsen JD, Hunter CN, Schulten K (Oct 2007). Atomic-level structural and functional model of a bacterial photosynthetic membrane vesicle . Proceedings of the National Academy of Sciences of the United States of America. 104 (40): 15723–8. Bibcode:2007PNAS..10415723S. doi:10.1073/pnas.0706861104. PMC 2000399 . PMID 17895378.
23. Campbell NA, Williamson B, Heyden RJ (2006). Biology Exploring Life. Upper Saddle River, NJ: Prentice Hall. ISBN 0-13-250882-6.
24. Raven PH, Evert RF, Eichhorn SE (2005). Biology of Plants, (7th ed.). New York: W. H. Freeman and Company. pp. 124–127. ISBN 0-7167-1007-2.
25. Yachandra Group Home page .
26. Pushkar Y, Yano J, Sauer K, Boussac A, Yachandra VK (Feb 2008). Structural changes in the Mn4Ca cluster and the mechanism of photosynthetic water splitting . Proceedings of the National Academy of Sciences of the United States of America. 105 (6): 1879–84. Bibcode:2008PNAS..105.1879P. doi:10.1073/pnas.0707092105. PMC 2542863 . PMID 18250316.
27. Williams BP, Johnston IG, Covshoff S, Hibberd JM (September 2013). Phenotypic landscape inference reveals multiple evolutionary paths to C4 photosynthesis . eLife. 2: e00961. doi:10.7554/eLife.00961. PMID 24082995.

28. Taiz L, Geiger E (2006). Plant Physiology (4th ed.). Sinauer Associates. ISBN 978-0-87893-856-8.
29. Monson RK, Sage RF (1999). The Taxonomic Distribution of C4 Photosynthesis . C_4 plant biology. Boston: Academic Press. pp. 551–580. ISBN 0-12-614440-0.
30. Dodd AN, Borland AM, Haslam RP, Griffiths H, Maxwell K (Apr 2002). "Crassulacean acid metabolism: plastic, fantastic". Journal of Experimental Botany. 53 (369): 569–80. doi:10.1093/jexbot/53.369.569. PMID 11886877.
31. Badger MR, Price GD (Feb 2003). "CO2 concentrating mechanisms in cyanobacteria: molecular components, their diversity and evolution". Journal of Experimental Botany. 54 (383): 609–22. doi:10.1093/jxb/erg076. PMID 12554704.
32. Badger MR, Andrews JT, Whitney SM, Ludwig M, Yellowlees DC, Leggat W, Price GD (1998). "The diversity and coevolution of Rubisco, plastids, pyrenoids, and chloroplast-based CO2-concentrating mechanisms in algae". Canadian Journal of Botany. 76 (6): 1052–1071. doi:10.1139/b98-074.
33. Miyamoto K. "Chapter 1 – Biological energy production". Renewable biological systems for alternative sustainable energy production (FAO Agricultural Services Bulletin – 128). Food and Agriculture Organization of the United Nations. Retrieved 2009-01-04.
34. Maxwell K, Johnson GN (Apr 2000). "Chlorophyll fluorescence--a practical guide". Journal of Experimental Botany. 51 (345): 659–68. doi:10.1093/jexbot/51.345.659. PMID 10938857.
35. Maxwell K., Johnson G. N, (2000) Chlorophyll fluorescence – a practical guide. Journal of Experimental Botany Vol. 51, No. 345, pp. 659-668- April 2000
36. Govindjee R. "What is Photosynthesis". Biology at Illinois.
37. Rosenqvist E., van Kooten O., (2006) Chlorophyll Fluorescence: A General Description and Nomenclature. From Chapter 2 "Practical Applications of Chlorophyll Fluorescence in Plant Biology". by Jennifer R. DeEll (Editor), Peter M.A. Toivonen (Editor) Kluwer Academic Publishers group, P.O Box 322, 3300 A.H. Dordrecht, the Netherlands, pages 39-78
38. Baker N. R., Oxborough K., (2004) Chlorophyll fluorescence as a probe of photosynthetic productivity. From Chapter 3, "Chlorophylla Fluorescence a Signature of Photosynthesis", edited by George Papaqeorgiou and Govindjee, published by Springer 2004, PO Box17, 3300 AA Dordrecht, The Netherlands, pages 66-79
39. Flexas 1999 – "Water stress induces different levels of photosynthesis and electron transport rate regulation in grapevines"J. FLEXAS, J. M. ESCALONA & H. MEDRANO Plant, Cell & Environment Volume 22 Issue 1 Page 39-48, January 1999
40. Fryer M. J., Andrews J.R., Oxborough K., Blowers D. A., Baker N.E. (1998) Relationship between CO2 assimilation, photosynthetic electron transport and active 02 metabolism in leaves of maize in the field during periods of low temperature

41. Earl H., Said Ennahli S., (2004) Estimating photosynthetic electron transport via chlorophyll fluorometry without Photosystem II light saturation. Photosynthesis Research 82: 177–186, 2004

42. Genty B., Briantais J. M. & Baker N. R. (1989) The relationship between the quantum yield of photosynthetic electron transport and quenching of chlorophyll fluorescence, Biochimica et Biophysica Acta 990, 87-92

43. Baker N. R. (2008) Chlorophyll Fluorescence: A Probe of PhotosynthesisIn Vivo Annu. Rev. Plant Biol. 2008. 59:89–113 The Annual Review of Plant Biology is online at plant.annualreviews.org, doi: 10.1146/annurev.arplant.59.032607.092759

44. Bernacchi C.J., Portis A. R., Nakano H., von Caemmerer S., and Long S.P. (2002) Temperature Response of Mesophyll Conductance. Implications for the Determination of Rubisco Enzyme Kinetics and for Limitations to Photosynthesis in Vivo Plant Physiology, December 2002, Vol. 130, pp. 1992–1998, http://www.plantphysiol.org © 2002 American Society of Plant Biologists

45. Ribas-Carbo M., Flexas J., Robinson S.A., Tcherkez G. G. B., (2010) In vivo measurement of plant respiration University of Wollongong Research Online

46. Long S.P., and Bernacchi C.J. (2003) Gas exchange measurements, what can they tell us about the underlying limitations to photosynthesis? Procedures and sources of error.

47. Ribas-Carbo M., Flexas J., Robinson S.A., Tcherkez G. G. B., (2010) In vivo measurement of plant respiration University of Wollongong Research Online

48. Bernacchi C.J., Portis A. R., Nakano H., von Caemmerer S., and Long S.P. (2002) Temperature Response of Mesophyll Conductance. Implications for the Determination of Rubisco Enzyme Kinetics and for Limitations to Photosynthesis in Vivo Plant Physiology, December 2002, Vol. 130, pp. 1992–1998, http://www.plantphysiol.org © 2002 American Society of Plant Biologists

49. YIN X., & STRUIK P.C., (2009) Theoretical reconsiderations when estimating the mesophyll conductanceto CO2 diffusion in leaves of C3 plants by analysis of combined gas exchange and chlorophyll fluorescence measurements pce_2016 1513..1 Plant, Cell and Environment (2009) 32, 1513–1524 524

50. Schreiber U, Klughammer C, Kolbowski J (2012). "Assessment of wavelength-dependent parameters of photosynthetic electron transport with a new type of multi-color PAM chlorophyll fluorometer". Photosynthesis research. 113 (1-3): 127–144. doi:10.1007/s11120-012-9758-1.

51. Palmer J (21 June 2013). "Plants 'seen doing quantum physics'". BBC News.

52. Lloyd S (10 March 2014). "Quantum Biology: Better Living Through Quantum Mechanics - The Nature of Reality". Nova: PBS Online, WGBH Boston.

53. Hildner R, Brinks D, Nieder JB, Cogdell RJ, van Hulst NF (Jun 2013). "Quantum coherent energy transfer over varying pathways in single light-harvesting complexes". Science. 340 (6139): 1448–51. doi:10.1126/science.1235820. PMID 23788794.

54. Photosynthesis got a really early start, New Scientist, 2 October 2004
55. Revealing the dawn of photosynthesis, New Scientist, 19 August 2006
56. Venn AA, Loram JE, Douglas AE (2008). "Photosynthetic symbioses in animals". Journal of Experimental Botany. 59 (5): 1069–80. doi:10.1093/jxb/erm328. PMID 18267943.
57. Rumpho ME, Summer EJ, Manhart JR (May 2000). "Solar-powered sea slugs. Mollusc/algal chloroplast symbiosis". Plant Physiology. 123 (1): 29–38. doi:10.1104/pp.123.1.29. PMC 1539252 . PMID 10806222.
58. Muscatine L, Greene RW (1973). "Chloroplasts and algae as symbionts in molluscs". International Review of Cytology. International Review of Cytology. 36: 137–69. doi:10.1016/S0074-7696(08)60217-X. ISBN 9780123643360. PMID 4587388.
59. Rumpho ME, Worful JM, Lee J, Kannan K, Tyler MS, Bhattacharya D, Moustafa A, Manhart JR (Nov 2008). "Horizontal gene transfer of the algal nuclear gene psbO to the photosynthetic sea slug Elysia chlorotica". Proceedings of the National Academy of Sciences of the United States of America. 105 (46): 17867–71. Bibcode:2008PNAS..10517867R. doi:10.1073/pnas.0804968105. PMC 2584685 . PMID 19004808.
60. Douglas SE (Dec 1998). "Plastid evolution: origins, diversity, trends". Current Opinion in Genetics & Development. 8 (6): 655–61. doi:10.1016/S0959-437X(98)80033-6. PMID 9914199.
61. Reyes-Prieto A, Weber AP, Bhattacharya D (2007). "The origin and establishment of the plastid in algae and plants". Annual Review of Genetics. 41: 147–68. doi:10.1146/annurev.genet.41.110306.130134. PMID 17600460.
62. Raven JA, Allen JF (2003). "Genomics and chloroplast evolution: what did cyanobacteria do for plants?". Genome Biology. 4 (3): 209. doi:10.1186/gb-2003-4-3-209. PMC 153454 . PMID 12620099.
63. Tomitani A, Knoll AH, Cavanaugh CM, Ohno T (Apr 2006). "The evolutionary diversification of cyanobacteria: molecular-phylogenetic and paleontological perspectives". Proceedings of the National Academy of Sciences of the United States of America. 103 (14): 5442–7. doi:10.1073/pnas.0600999103. PMC 1459374 . PMID 16569695.
64. "Cyanobacteria: Fossil Record". Ucmp.berkeley.edu. Retrieved 2010-08-26.
65. Smith, Alison (2010). Plant biology. New York, NY: Garland Science. p. 5. ISBN 0815340257.
66. Herrero A, Flores E (2008). The Cyanobacteria: Molecular Biology, Genomics and Evolution (1st ed.). Caister Academic Press. ISBN 978-1-904455-15-8.
67. Walker DA (2002). "'And whose bright presence' - an appreciation of Robert Hill and his reaction" (PDF). Photosynthesis Research. 73 (1-3): 51–4. doi:10.1023/A:1020479620680. PMID 16245102.

68. Otto Warburg – Biography. Nobelprize.org (1970-08-01). Retrieved on 2011-11-03.
69. Gest, Howard (2002). "History of the word photosynthesis and evolution of its definition.". Photosynthesis Research. 73 (1-3): 7–10. doi:10.1023/A:1020419417954.
70. Calvin, Melvin (July 1989). "Forty years of photosynthesis and related activities". Photosynthesis Research. 21 (1). doi:10.1007/BF00047170.
71. Verduin, J. 1953. A table of photosynthesis rates under optimal, near natural conditions. Am.J. Bot. 40:675-679.
72. Verduin, J., Whitwer, E. E., and Cowell, B.C. 1959. Maximal photosynthetic rates in nature, Science 130:268-269.
73. Hesketh, J. D., and Musgrave, R.B. 1962. Photosynthesis under field conditions. IV. Light studies with individual corn leaves. Crop Sci. 2:311-315.
74. Hesketh, J.D., and Moss, D. N. 1963.Variation in the response of photosynthesis to light. Crop Sci. 3:107-110.
75. El-Sharkawy, M. A., and Hesketh, J. D. 1965. Photosynthesis among species in relation to characteristics of leaf anatomy and CO2 diffusion resistances. Crop Sci. 5:517-521.
76. El-Sharkawy, M. A., and Hesketh, J. D. 1986.Citation Classic-Photosynthesis among species in relation to characteristics of leaf anatomy and CO2 diffusion resistances. Curr. Cont./Agr.Biol.Environ.27:14-14. Online http://www.library.upenn/edu/classics1986/A1986C691300001.pdf.
77. Haberlandt, G. 1904.Physiologische Pflanzanatomie. Engelmann, Leipzig.
78. El-Sharkawy, M. A. 1965.Factors Limiting Photosynthetic Rates of Different Plant Species. Ph.D. Dissertation, The University of Arizona, Tucson, USA.
79. Karpilov, Y.S. 1960. The distribution of radioactvity in carbon-14 among the products of photosynthesis in maize. Proc. Kazan Agric. Inst. 14:15-24.
80. Kortschak, H.P., Hart, C.E., and Burr, G.O. 1965. Carbon dioxide fixation in sugarcane leaves. Plant Physiol. 40:209-213.
81. Hatch, M.D., and Slack, C. R. 1966.Photosynthesis by sugar-cane leaves. A new carboxylation reaction and the pathway of sugar formation. Biochem.J. 101:103-111.
82. Jones HG (2014). Plants and Microclimate: a Quantitative Approach to Environmental Plant Physiology (Third ed.). Cambridge: Cambridge University Press. ISBN 978-0-521-27959-8.

CHAPTER 2

Chlorophyll Function and Chemistry

Chapter Contents

Chlorophyll is a complex organic molecule that enables plants, and some other organisms, to carry out photosynthesis — the conversion of carbon dioxide and water to glucose and oxygen using sunlight. The glucose is a source of energy that is used to power growth and development, and the oxygen created as a byproduct is essential to all animal life. As a result, this green pigment is one of the most important biological compounds. It also gives plants their distinctive green color.

INTRODUCTION

Chlorophyll is a term used for several closely related green pigments found in cyanobacteria and the chloroplasts of algae and plants. Its name is derived from the Greek words χλωρός, chloros ("green") and φύλλον, phyllon ("leaf"). Chlorophyll is essential in photosynthesis, allowing plants to absorb energy from light. Chlorophyll absorbs light most strongly in the blue portion of the electromagnetic spectrum, followed by the red portion. Conversely, it is a poor absorber of green and near-green portions of the spectrum, which it reflects, producing the green color of chlorophyll-containing tissues. Chlorophyll molecules are specifically arranged in and around photosystems that are embedded in the thylakoid membranes of chloroplasts. Two types of chlorophyll exist in the photosystems of green plants: chlorophyll a and b. Chlorophyll was first isolated and named by Joseph Bienaimé Caventou and Pierre Joseph Pelletier in 1817.

Chlorophyll is responsible for the green color of many plants and algae.

CHLOROPHYLL AND PHOTOSYNTHESIS

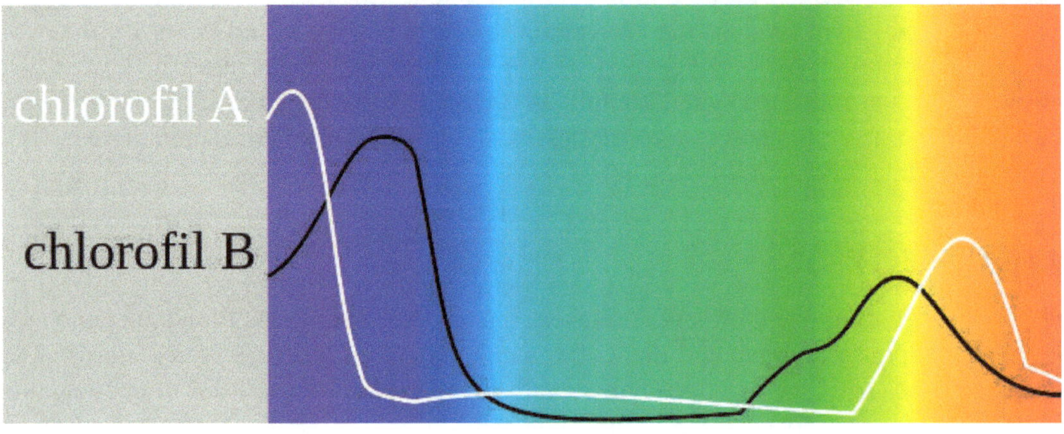

Absorption maxima of chlorophylls against the spectrum of white light. Chlorophyll absorbs energy from light strongly at visible frequencies.

Chlorophyll is vital for photosynthesis, which allows plants to absorb energy from light.

Chlorophyll molecules are specifically arranged in and around photosystems that are embedded in the thylakoid membranes of chloroplasts. In these complexes, chlorophyll serves two primary functions. The function of the vast majority of chlorophyll (up to several hundred molecules per photosystem) is to absorb light and transfer that light energy by resonance energy transfer to a specific chlorophyll pair in the reaction center of the photosystems. The two currently accepted photosystem units are photosystem II and photosystem I, which have their own distinct reaction centres, named P680 and P700, respectively. These centres are named after the wavelength (in nanometers) of their red-peak absorption maximum. The identity, function and spectral properties of the types of chlorophyll in each photosystem are distinct and determined by each other and the protein structure surrounding them. Once extracted from the protein into a solvent (such as acetone or methanol), these chlorophyll pigments can be separated into chlorophyll a and chlorophyll b.

The function of the reaction center of chlorophyll is to absorb light energy and transfer it to other parts of the photosystem. The absorbed energy of the photon is transferred to an electron in a process called charge separation. The removal of the electron from the chlorophyll is an oxidation reaction. The chlorophyll donates the high energy electron to a series of molecular intermediates called an electron transport chain. The charged reaction center of chlorophyll ($P680^+$) is then reduced back to its ground state by accepting an electron stripped from water. The electron that reduces $P680^+$ ultimately comes from the oxidation of water into $O_2$and H^+ through several intermediates. This reaction is how photosynthetic organisms such as plants produce O_2 gas, and is the source for practically all the O_2 in Earth's atmosphere. Photosystem I typically works in series with Photosystem II; thus the $P700^+$ of Photosystem I is usually reduced as it accepts the electron, via many intermediates in the thylakoid membrane, by electrons coming, ultimately, from Photosystem II. Electron transfer reactions in the thylakoid membranes are complex, however, and the source of electrons used to reduce $P700^+$ can vary.

The electron flow produced by the reaction center chlorophyll pigments is used to pump H^+ ions across the thylakoid membrane, setting up a chemiosmotic potential used mainly in the production of ATP (stored chemical energy) or to reduce $NADP^+$ to NADPH. NADPH is a universal agent used to reduce CO_2 into sugars as well as other biosynthetic reactions.

Reaction center chlorophyll–protein complexes are capable of directly absorbing light and performing charge separation events without other the assistance of other chlorophyll pigments, but the probability of that happening under a given light intensity is small. Thus, the other chlorophylls in the photosystem and antenna pigment proteins all cooperatively absorb and funnel light energy to the reaction center. Besides chlorophyll *a*, there are other pigments, called accessory pigments, which occur in these pigment–protein antenna complexes.

CHEMICAL STRUCTURE

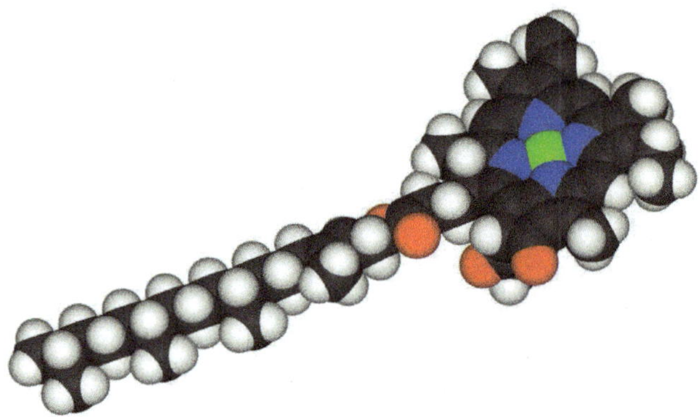

Space-filling model of the chlorophyll *a* molecule

Chlorophyll is a chlorin pigment, which is structurally similar to and produced through the same metabolic pathway as other porphyrin pigments such as heme. At the center of the chlorin ring is a magnesium ion. This was discovered in 1906, and was the first time that magnesium had been detected in living tissue. For the structures depicted in this article, some of the ligands attached to the Mg^{2+} center are omitted for clarity. The chlorin ring can have several different side chains, usually including a long phytol chain. There are a few different forms that occur naturally, but the most widely distributed form in terrestrial plants is chlorophyll *a*. After initial work done by German chemist Richard Willstätter spanning from 1905 to 1915, the general structure of chlorophyll *a* was elucidated by Hans Fischer in 1940. By 1960, when most of the stereochemistry of chlorophyll *a* was known, Robert Burns Woodward published a total synthesis of the molecule. In 1967, the last remaining stereochemical elucidation was completed by Ian Fleming, and in 1990 Woodward and co-authors published an updated synthesis. Chlorophyll f was announced to be present in cyanobacteria and other oxygenic microorganisms that form stromatolites in 2010; a molecular formula of $C_{55}H_{70}O_6N_4Mg$ and a structure of (2-formyl)-chlorophyll *a* were deduced based on NMR, optical and mass spectra. The different structures of chlorophyll are summarized below:

	Chlorophyll a	Chlorophyll b	Chlorophyll c1	Chlorophyll c2	Chlorophyll d	Chlorophyll f
Molecular formula	$C_{55}H_{72}O_5N_4Mg$	$C_{55}H_{70}O_6N_4Mg$	$C_{35}H_{30}O_5N_4Mg$	$C_{35}H_{28}O_5N_4Mg$	$C_{54}H_{70}O_6N_4Mg$	$C_{55}H_{70}O_6N_4Mg$
C2 group	$-CH_3$	$-CH_3$	$-CH_3$	$-CH_3$	$-CH_3$	-CHO
C3 group	$-CH{=}CH_2$	$-CH{=}CH_2$	$-CH{=}CH_2$	$-CH{=}CH_2$	-CHO	$-CH{=}CH_2$
C7 group	$-CH_3$	-CHO	$-CH_3$	$-CH_3$	$-CH_3$	$-CH_3$
C8 group	$-CH_2CH_3$	$-CH_2CH_3$	$-CH_2CH_3$	$-CH{=}CH_2$	$-CH_2CH_3$	$-CH_2CH_3$
C17 group	$-CH_2CH_2COO-$ Phytyl	$-CH_2CH_2COO-$ Phytyl	-CH=CHCOOH	-CH=CHCOOH	$-CH_2CH_2COO-$ Phytyl	$-CH_2CH_2COO-$ Phytyl
C17-C18 bond	Single(chlorin)	Single (chlorin)	Double (porphyrin)	Double (porphyrin)	Single (chlorin)	Single (chlorin)
Occurrence	Universal	Mostly plants	Various algae	Various algae	Cyanobacteria	Cyanobacteria

When leaves degreen in the process of plant senescence, chlorophyll is converted to a group of colourless tetrapyrroles known as nonfluorescent chlorophyll catabolites (NCC's) with the general structure:

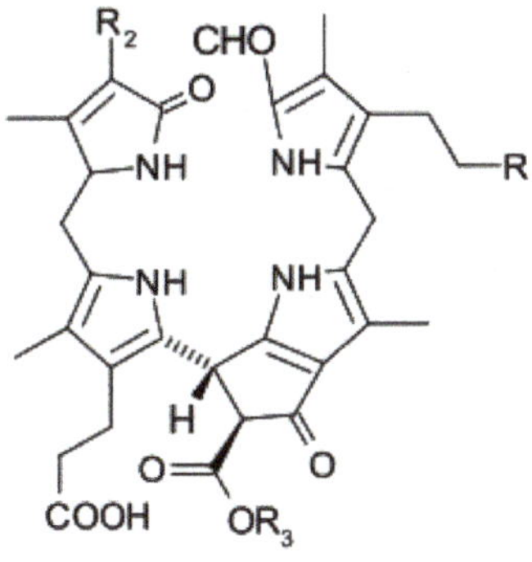

These compounds have also been identified in several ripening fruits.

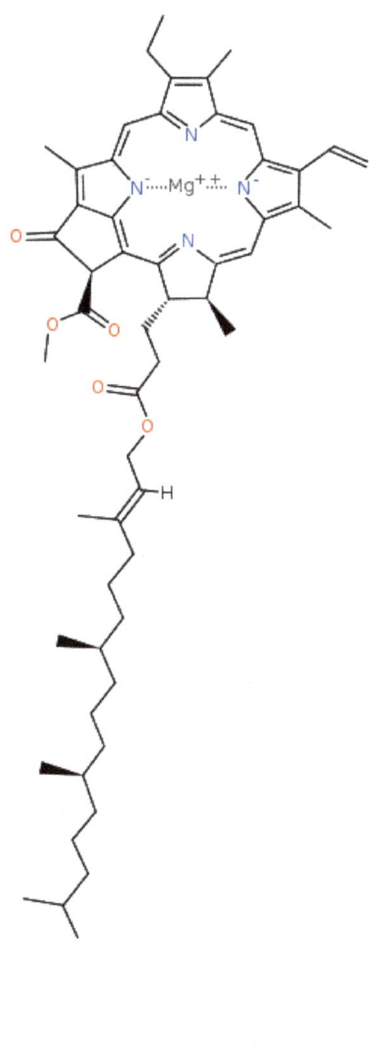

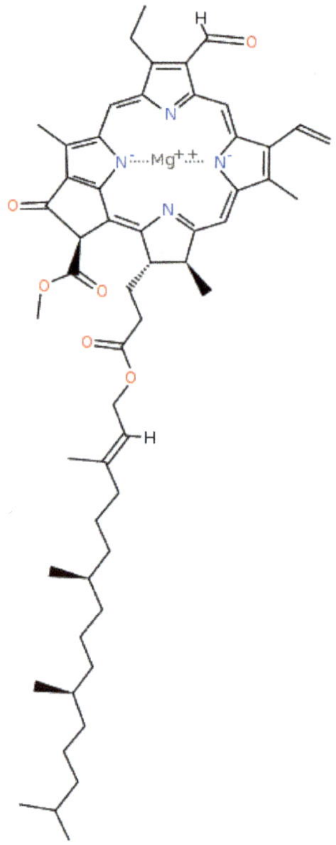
H
O
N
N⁻····Mg⁺⁺····N⁻
N
O
O
O
O
O
H

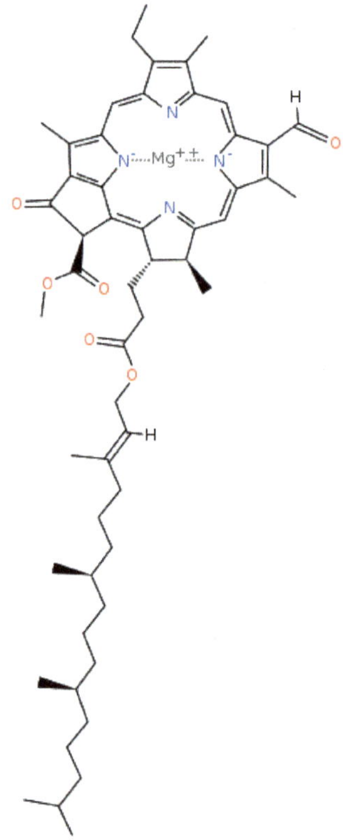
H
O
N
N⁻····Mg⁺⁺····N⁻
N
O
O
O
O
O
H

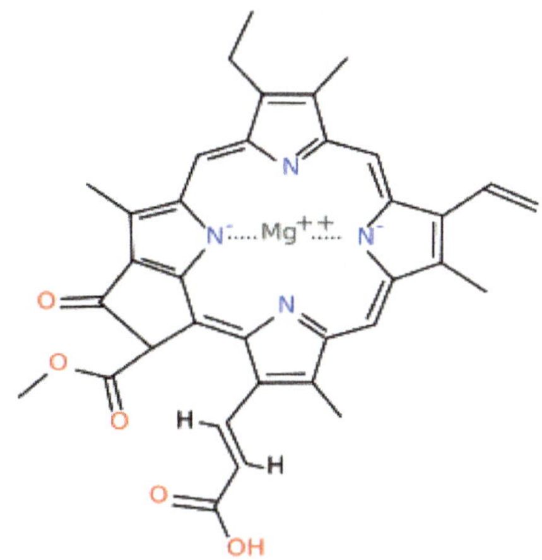
N
N⁻·····Mg⁺⁺····· N⁻
N
O
O
O
H
H
O
OH

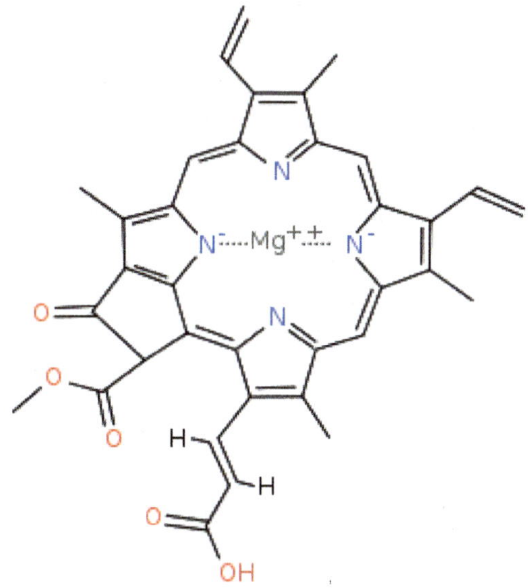
N
N⁻·····Mg⁺⁺····· N⁻
N
O
O
O
H
H
O
OH

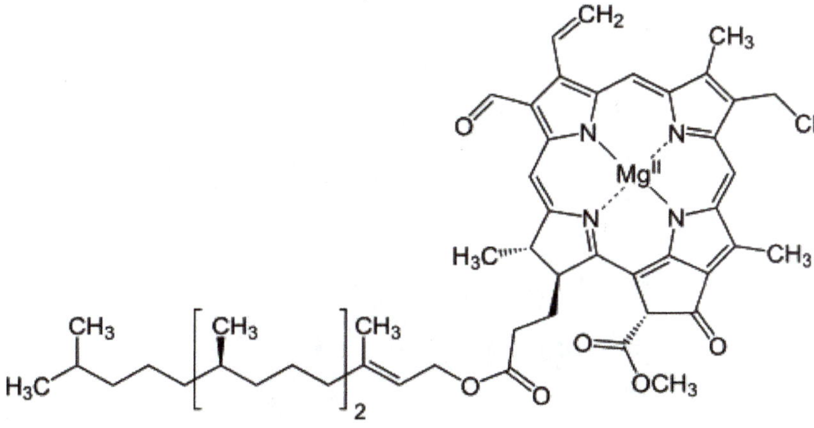
CH₂
CH₃
O
CH₃
N
N
Mgᴵᴵ
N
N
H₃C
CH₃
CH₃
CH₃
CH₃
H₃C
2
O
O
O
O
OCH₃

Measurement of Chlorophyll Content

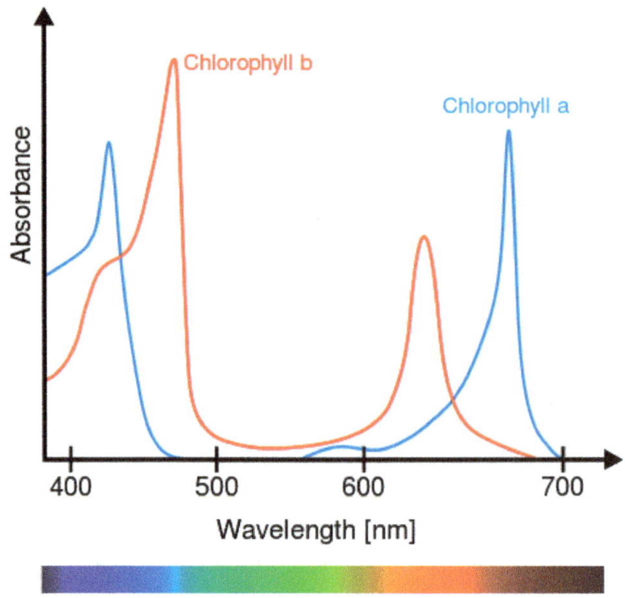

Absorbance spectra of free chlorophyll *a* (blue) and *b* (red) in a solvent. The spectra of chlorophyll molecules are slightly modified *in vivo* depending on specific pigment-protein interactions.

Because it absorbs red and blue light strongly but is transparent to green light, pure chlorophyll has a strong green colour.

Measurement of the absorption of light is complicated by the solvent used to extract the chlorophyll from plant material, which affects the values obtained,

- In diethyl ether, chlorophyll *a* has approximate absorbance maxima of 430 nm and 662 nm, while chlorophyll *b* has approximate maxima of 453 nm and 642 nm.
- The absorption peaks of chlorophyll *a* are at 665 nm and 465 nm. Chlorophyll *a* fluoresces at 673 nm (maximum) and 726 nm. The peak molar absorption coefficient of chlorophyll *a* exceeds 10^5 M^{-1} cm^{-1}, which is among the highest for small-molecule organic compounds.
- In 90% acetone-water, the peak absorption wavelengths of chlorophyll *a* are 430 nm and 664 nm; peaks for chlorophyll *b* are 460 nm and 647 nm; peaks for chlorophyll *c1* are 442 nm and 630 nm; peaks for chlorophyll *c2* are 444 nm and 630 nm; peaks for chlorophyll *d* are 401 nm, 455 nm and 696 nm.

By measuring the absorption of light in the red and far red regions, it is possible to estimate the concentration of chlorophyll within a leaf.

Ratio fluorescence emission can be used to measure chlorophyll content. By exciting chlorophyll "a" fluorescence at a lower wavelength, the ratio of chlorophyll fluorescence emission at 705 nm +/- 10 nm and 735 nm +/-10 nm can provide a linear relationship of chlorophyll content when compared to chemical testing. The ratio F735/F700 provided a correlation value of r^2 0.96 compared to chemical testing in the range from 41 mg m^{-2} up to 675 mg m^{-2}. Gitelson also developed a formula for direct readout of chlorophyll content in mg m^{-2}. The formula provided a reliable method of measuring chlorophyll content from 41 mg m^{-2} up to 675 mg m^{-2} with a correlation r^2 value of 0.95.

Biosynthesis

In plants, chlorophyll may be synthesized from succinyl-CoA and glycine, although the immediate precursor to chlorophyll *a* and *b* is protochlorophyllide. In Angiosperm plants, the last step, the conversion of protochlorophyllide to chlorophyll, is light-dependent and such plants are pale (etiolated) if grown in darkness. Non-vascular plants and green algae have an additional light-independent enzyme and grow green even in darkness.

Chlorophyll itself is bound to proteins and can transfer the absorbed energy in the required direction. Protochlorophyllide occurs mostly in the free form and, under light conditions, acts as a photosensitizer, forming highly toxic free radicals. Hence, plants need an efficient mechanism of regulating the amount of chlorophyll precursor. In angiosperms, this is done at the step of aminolevulinic acid (ALA), one of the intermediate compounds in the biosynthesis pathway. Plants that are fed by ALA accumulate high and toxic levels of protochlorophyllide; so do the mutants with the damaged regulatory system.

Chlorosis is a condition in which leaves produce insufficient chlorophyll, turning them yellow. Chlorosis can be caused by a nutrient deficiency of iron—called iron chlorosis—or by a shortage of magnesium or nitrogen. Soil pH sometimes plays a role in nutrient-caused chlorosis; many plants are adapted to grow in soils with specific pH levels and their ability to absorb nutrients from the soil can be dependent on this. Chlorosis can also be caused by pathogens including viruses, bacteria and fungal infections, or sap-sucking insects.

Complementary Light Absorbance of Anthocyanins with Chlorophylls

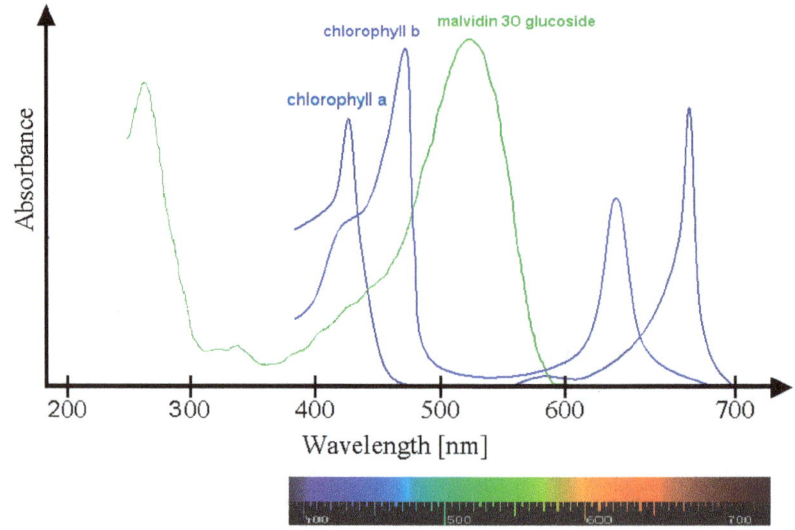

Superposition of spectra of chlorophyll a and b with oenin (malvidin 3O glucoside), a typical anthocyanidin, showing that, while chlorophylls absorb in the blue and yellow/red parts of the visible spectrum, oenin absorbs mainly in the green part of the spectrum, where chlorophylls don›t absorb at all.

Anthocyanins are other plant pigments. The absorbance pattern responsible for the red color of anthocyanins may be complementary to that of green chlorophyll in photosynthetically active tissues such as young *Quercus coccifera* leaves. It may protect the leaves from attacks by plant eaters that may be attracted by green color.

Distribution

Animation depicting nearly four years worth of SeaWiFS ocean chlorophyll concentration.

The chlorophyll maps show milligrams of chlorophyll per cubic meter of seawater each month. Places where chlorophyll amounts were very low, indicating very low numbers of phytoplankton, are blue. Places where chlorophyll concentrations were high, meaning many phytoplankton were growing, are yellow. The observations come from the Moderate Resolution Imaging Spectroradiometer (MODIS) on NASA›s Aqua satellite. Land is dark gray, and places where MODIS could not collect data because of sea ice, polar darkness, or clouds are light gray.The highest chlorophyll concentrations, where tiny surface-dwelling ocean plants are thriving, are in cold polar waters or in places where ocean currents bring cold water to the surface, such as around the equator and along the shores of continents. It is not the cold water itself that stimulates the phytoplankton. Instead, the cool temperatures are often a sign that the water has welled up to the surface from deeper in the ocean, carrying nutrients that have built up over time. In polar waters, nutrients accumulate in surface waters during the dark winter months when plants cannot grow. When sunlight returns in the spring and summer, the plants flourish in high concentrations.

Culinary Use

Chlorophyll is registered as a food additive (colorant), and its E number is E140. Chefs use chlorophyll to color a variety of foods and beverages green, such as pasta and absinthe. Chlorophyll is not soluble in water, and it is first mixed with a small quantity of vegetable oil to obtain the desired solution.

CHLOROPHYLL FLUORESCENCE

Chlorophyll fluorescence is light re-emitted by chlorophyll molecules during return from excited to non-excited states and used as indicator of photosynthetic energy conversion in higher plants, algae and bacteria. Excited chlorophyll dissipates the absorbed light energy by driving photosynthesis (photochemical energy conversion), as heat in non-photochemical quenching or by emission as fluorescence radiation. As these processes are complementary processes analysis of chlorophyll fluorescence is an important tool in plant research with a wide spectra of applications.

A chlorophyll extract in alcohol shown under white light (above) and UV light inducing fluorescence (below).

Measuring Fluorescence

Usually the initial measurement is the minimal level of fluorescence,F_0This is the fluorescence in the absence of photosynthetic light.

To use measurements of chlorophyll fluorescence to analyse photosynthesis, researchers must distinguish between photochemical quenching and non-photochemical quenching (heat dissipation). This is achieved by stopping photochemistry, which allows researchers to measure fluorescence in the presence of non-photochemical quenching alone. To reduce photochemical quenching to negligible levels, a high intensity, short flash of light is applied to the leaf. This transiently closes all PSII reaction centres, which prevents energy of PSII being passed to downstream electron carriers. Non-photochemical quenching will not be affected if the flash is short. During the flash, the fluorescence reaches the level reached in the absence of any photochemical quenching, known as maximum fluorescence F_m

The efficiency of photochemical quenching (which is a proxy of the efficiency of PSII) can be estimated by comparing F_m to the steady yield of fluorescence in the light F_m to the steady yield of fluorescence in the light F_t. The efficiency of non-photochemical quenching is altered by various internal and external factors. Alterations in heat dissipation mean changes in F_m Heat dissipation cannot be totally stopped, so the yield of chlorophyll fluorescence in the absence of non-photochemical quenching cannot be measured. Therefore, researchers use a dark-adapted point (F_m^0) with which to compare estimations of non-photochemical quenching.

Common Fluorescence Parameters

F_0 Minimal fluorescence (arbitrary units). Fluorescence level when all antenna pigment complexes associated with the photosystem are assumed to be open (dark adapted).

F_m : Maximal fluorescence (arbitrary units). Fluorescence level when a high intensity flash has been applied. All antenna sites are assumed to be closed.

F_0': Minimal fluorescence (arbitrary units). Fluorescence level of illuminated sample which is lowered with respect to F_0 by non-photochemical quenching.

F_m' : Maximal fluorescence (arbitrary units). Fluorescence level of illuminated sample as induced by saturating pulses which temporarily close all PSII reaction centers.

F_{tr} : Terminal fluorescence (arbitrary units). Fluorescence quenching value at the end of the test.

$T_{1/2}$: Half rise time from F_0 to F_m.

Calculated Parameters

F_v is variable fluorescence. Calculated as $F_v = F_m$. $\frac{F_v}{F_m}$ is the ratio of variable fluorescence to maximal fluorescence. Calculated as $\frac{F_m - F_0}{F_m}$. This is a measure of the maximum efficiency of PSII (the efficiency if all PSII centres were open). $\frac{F_v}{F_m}$ can be used to estimate the potential efficiency of PSII by taking dark-adapted measurements. Φ_{PSII} measures the efficiency of Photosystem II. Calculated as $Y_{(II)} = , \frac{F_m' - F}{F_m'}$. This parameter measures the proportion of light absorbed by PSII that is used in photochemistry. As such, it can give a measure of the rate of linear electron transport and so indicates overall photosynthesis.

qP (photochemical quenching). Calculated as $\frac{F_m' - F}{F_m' - F_0'}$ This parameter approximates the proportion of PSII reaction centres that are open.

Whilst Φ_{PSII} gives an estimation of the efficiency, qP and $\frac{F_v}{F_m}$ tell us which processes which have altered the efficiency. Closure of reaction centers as a result of a high intensity light will alter the value of qP. Changes in the efficiency of non-photochemical quenching will alter the ratio $\frac{F_v}{F_m}$.

Applications of the Theory

PSII Yield as a Measure of Photosynthesis

Chlorophyll fluorescence appears to measure of photosynthesis, but this is an over-simplification. Fluorescence can measure the efficiency of PSII photochemistry, which can be used to estimate the rate of linear electron transport by multiplying by the light intensity. However, researchers generally mean carbon fixation when they refer to photosynthesis. Electron transport and CO_2 fixation can correlate well, but may not correlate in the field due to processes such as photorespiration, nitrogen metabolism and the Mehler reaction.

Relating Electron Transport to Carbon Fixation

A powerful research technique is to simultaneously measure chlorophyll fluorescence and gas exchange to obtain a full picture of the response of plants to their environment. One technique is to simultaneously measure CO_2 fixation and PSII photochemistry at different light intensities, in non-photorespiratory conditions. A plot of CO_2 fixation and PSII photochemistry indicates the electron requirement per molecule CO_2 fixed. From this estimation, the extent of photorespiration may be estimated. This has been used to explore the significance of photorespiration as a photoprotective mechanism during drought.

Fluorescence analysis can also be applied to understanding the effects of low and high temperatures.

- Sobrado (2008) investigated gas exchange and chlorophyll *a* fluorescence responses to high intensity light, of pioneer species and forest species. Midday leaf gas exchange was measured using a photosynthesis system, which measured net photosynthetic rate, gs, and intercellular CO_2 concentration (C_i) In the same leaves used for gas exchange measurements, chlorophyll *a* fluorescence parameters were measured using a fluorometer. The results showed that despite pioneer species and forest species occupying different habitats, both showed similar vulnerability to midday photoinhibition in sun-exposed leaves.

Measuring Stress and Stress Tolerance

Chlorophyll fluorescence can measure most types of plant stress. Chlorophyll fluorescence can be used as a proxy of plant stress because environmental stresses, e.g. extremes of temperature, light and water availability, can reduce the ability of a plant to metabolise normally. This can mean an imbalance between the absorption of light energy by chlorophyll and the use of energy in photosynthesis.

- Favaretto et al. (2010) investigated adaptation to a strong light environment in pioneer and late successional species, grown under 100% and 10% light. Numerous parameters, including chlorophyll *a* fluorescence, were measured. A greater decline in $\frac{F_v}{F_m}$
- under full sun light in the late-successional species than in the pioneer species was observed. Overall, their results show that pioneer species perform better under high-sun light than late- successional species, suggesting that pioneer plants have more potential tolerance to photo-oxidative damage.
- Neocleous and Vasilakakis (2009) investigated the response of raspberry to boron and salt stress. An chlorophyll fluorometer was used to measure F_0, F_m and F_v
- The leaf chlorophyll fluorescence was not significantly affected by NaCl concentration when B concentration was low. When B was increased, leaf chlorophyll fluorescence was reduced under saline conditions. It could be concluded that the combined effect of B and NaCl on raspberries induces a toxic effect in photochemical parameters.
- Lu and Zhang (1999) studied heat stress in wheat plants and found that temperature stability in the Photosystem II of water-stressed leaves correlates positively to the resistance in metabolism during photosynthesis.

Nitrogen Balance Index

Example of a portable multiparametric fluorometer that uses the ratio between chlorophyll and flavonols to detect the nitrogen deficiency of plants

Because of the link between chlorophyll content and nitrogen content in leaves, chlorophyll fluorometers can be used to detect nitrogen deficiency in plants, by several methods.

Based on several years of research and experimentation, polyphenols can be the indicators of nitrogen status of a plant. For instance, when a plant is under optimal conditions, it favours its primary metabolism and synthesises the proteins (nitrogen molecules) containing chlorophyll, and few flavonols (carbon-based secondary compounds). On the other hand, in case of lack of nitrogen, we will observe an increased production of flavonols by the plant.

The NBI (Nitrogen Balance Index) by Force-A, allows the assessment of nitrogen conditions of a culture by calculating the ratio between Chlorophyll and Flavonols (related to Nitrogen/Carbon allocation) .

Measure Chlorophyll Content

Gitelson (1999) states, "The ratio between chlorophyll fluorescence at 735 nm and the wavelength range 700nm to 710 nm, F735/F700 was found to be linearly proportional to the chlorophyll content (with determination coefficient, r2, more than 0.95) and thus this ratio can be used as a precise indicator of chlorophyll content in plant leaves."

Chlorophyll Fluorometers

Fluorescence image (Ft value) of adaxial leaf surface

The development of fluorometers allowed chlorophyll fluorescence analysis to become a common method in plant research. Chlorophyll fluorescence analysis has been revolutionized by the invention of the Pulse-Amplitude-Modulation (PAM) technique and availability of the first commercial modulated chlorophyll fluorometer PAM-101 (Walz, Germany). By modulating the measuring light beam (microsecond-range pulses) and parallel detection of the excited fluorescence the relative fluorescence yield (Ft) can be determined in the presence of ambient light. Crucially, this means chlorophyll fluorescence can be measured in the field even in full sunlight.

Today, chlorophyll fluorometers are designed for measuring many different plant mechanisms. The measuring protocols: F_V/F_M and OJIP measure the efficiency of Photosystem II samples at a common and know dark adapted state. These protocols are useful in measuring many types of plant stress. Bernard Genty's light adapted measuring protocol $\Delta F/F_M$', or Y(II), is an effective and sensitive way to measure plant samples under ambient or artificial lighting conditions.However, since Y(II) values also change with light intensity, one should compare samples at the same light intensity unless light stress is the focus of the measurement. Y(II) can be more sensitive to some types of plant stress than F_V/F_M, such as heat stress.

Other plant mechanism measuring protocols have also been developed. When a chloroplast absorbs light, some of the light energy goes to photochemistry, some goes to regulated heat dissipation, and some goes to unregulated heat dissipation. Various chlorophyll fluorescence measuring parameters exist to measure all of these events. In the lake model, q_L measures photochemical quenching, Y(NYO) measures plant regulated heat dissipation, and Y(NO) measures unregulated heat dissipation. An older quenching protocol, called the puddle model, uses q_P for photochemical quenching, q_N for nonphotochemical quenching of both regulated and unregulated heat dissipation and NPQ for an estimate of nonphotochemical quenching. NPQ has also been resurrected to the lake model mathematically.

In addition, the parameters q_E, and pNPQ have been developed to measure the photoprotective xanthophyll cycle. q_T is a measure of state transitions. $q_M$is a measure of chloroplast migration, and q_I is a measure of plant photoinhibition.

At lower actinic light levels NPQ = qE+qT+qI

At high actinic light levels NPQ = qE+qM=qI

Some fluorometers are designed to be portable and operated in one hand.

Consistent further development into imaging fluorometers facilitate the visualization of spatial heterogeneities in photosynthetic activity of samples. These heterogeneities naturally occur in plant leaves for example during growths, various environmental stresses or pathogen infection. Thus knowledge about sample heterogeneities is important for correct interpretation of the photosynthetic performance of the plant sample. High performance imaging fluorometer systems provide options to analyze single cell/single chloroplast as well as sample areas covering whole leaves or plants.

Alternative Approaches

LIF Sensors

Techniques based on the Kautsky effect do not exhaust the variety of detection and evaluation methods based on the chlorophyll fluorescence. In particular, recent advances in the area of laser-induced fluorescence (LIF) also provide an opportunity of developing sufficiently compact and efficient sensors for photophysiological status and biomass assessments. Instead of measuring the evolution of the total fluorescence flux, such sensors record the spectral density of this flux excited by strong monochromatic laser light pulses of nanoseconds duration. Requiring no 15- 20 min dark adaptation period (as is the case for the Kautsky effect methods) and being capable to excite the sample from considerable distance, the LIF sensors can provide fast and remote evaluation.

- Application of the LIF technique to the assessment of drought stress in cork oak (*Quercus suber*) and maritime pine (*Pinus pinaster*) on the basis of chlorophyll emission ratio I_{685}/I_{740} is described in Ref. Recently the LIF sensing technique was harnessed to address the role of pPLAIIα protein in the protection of the photosynthetic metabolism during drought stress using genetically modified Arabidopsis plants.
- In 2011, Vieira et al. applied a compact low-cost LIF sensor (built around a frequency-doubled solid-state Q-switched Nd:YAG laser and a specially modified commercial miniature fiber optic spectrometer Ocean Optics USB4000) to study intertidal microphytobenthos communities. Chlorophyll emission enabled the researchers to adequately assess the surface biomass and track migratory rhythms of epipelic benthic microalgae in muddy sediments.

REFERENCES

1. May, Paul. "Chlorophyll". University of Bristol.
2. "chlorophyll". Online Etymology Dictionary.
3. Speer, Brian R. (1997). "Photosynthetic Pigments". UCMP Glossary (online). University of California Museum of Paleontology. Retrieved 2010-07-17.
4. See:
5. Delépine, Marcel (September 1951). "Joseph Pelletier and Joseph Caventou". Journal of Chemical Education. 28 (9): 454. Bibcode:1951JChEd..28..454D. doi:10.1021/ed028p454.
6. Pelletier and Caventou (1817) "Notice sur la matière verte des feuilles"(Notice on the green material in leaves), Journal de Pharmacie, 3 : 486-491. On p. 490, the authors propose a new name for chlorophyll. From p. 490: "Nous n'avons aucun droit pour nommer une substance connue depuis long-temps, et à l'histoire de laquelle nous n'avons ajouté que quelques faits ; cependant nous proposerons, sans y mettre aucune importance, le nom de chlorophyle, de chloros, couleur, et φυλλον, feuille : ce nom indiquerait le rôle qu›elle joue dans la nature." (We have no right to name a substance that has been known for a long time, and to whose story we have added only a few facts ; however, we will propose, without giving it any importance, the name chlorophyll, from chloros, color, and φυλλον, leaf : this name would indicate the role that it plays in nature.)
7. Carter, J. Stein (1996). "Photosynthesis". University of Cincinnati.
8. Nature (July 5, 2013). "Unit 1.3. Photosynthetic Cells". Essentials of Cell Biology. nature.com.

9. Marker, A. F. H. (1972). "The use of acetone and methanol in the estimation of chlorophyll in the presence of phaeophytin". Freshwater Biology. 2 (4): 361–385. doi:10.1111/j.1365-2427.1972.tb00377.x.
10. Jeffrey, S. W.; Shibata, Kazuo (February 1969). "Some Spectral Characteristics of Chlorophyll c from Tridacna crocea Zooxanthellae". Biological Bulletin. Marine Biological Laboratory. 136 (1): 54–62. doi:10.2307/1539668. JSTOR 1539668.
11. Gilpin, Linda (21 March 2001). "Methods for analysis of benthic photosynthetic pigment". School of Life Sciences, Napier University. Archived from the originalon April 14, 2008. Retrieved 2010-07-17.
12. Willstätter, Richard (1906) "Zur Kenntniss der Zusammensetzung des Chlorophylls"
13. Motilva, Maria-José (2008). "Chlorophylls – from functionality in food to health relevance". 5th Pigments in Food congress- for quality and health (Print). University of Helsinki. ISBN 978-952-10-4846-3.
14. Woodward, R. B.; Ayer, W. A.; Beaton, J. M.; Bickelhaupt, F.; Bonnett, R.; Buchschacher, P.; Closs, G. L.; Dutler, H.; Hannah, J.; et al. (July 1960). "The total synthesis of chlorophyll". Journal of the American Chemical Society. 82 (14): 3800–3802. doi:10.1021/ja01499a093.
15. Fleming, Ian (14 October 1967). "Absolute Configuration and the Structure of Chlorophyll". Nature. 216 (5111): 151–152. Bibcode:1967Natur.216..151F. doi:10.1038/216151a0.
16. Woodward, R. B.; Ayer, William A.; Beaton, John M.; Bickelhaupt, Friedrich; Bonnett, Raymond; Buchschacher, Paul; Closs, Gerhard L.; Dutler, Hans; Hannah, John; et al. (1990). "The total synthesis of chlorophyll a" (PDF). Tetrahedron. 46 (22): 7599–7659. doi:10.1016/0040-4020(90)80003-Z.
17. Jabr, Ferris (August 19, 2010) A New Form of Chlorophyll?. Scientific American. Retrieved on 2012-04-15.
18. Infrared chlorophyll could boost solar cells. New Scientist. August 19, 2010. Retrieved on 2012-04-15.
19. Chen, Min; Schliep, Martin; Willows, Robert D.; Cai, Zheng-Li; Neilan, Brett A.; Scheer, Hugo (September 2010). "A Red-Shifted Chlorophyll". Science. 329 (5997): 1318–1319. Bibcode:2010Sci...329.1318C. doi:10.1126/science.1191127. PMID 20724585.
20. Müller, Thomas; Ulrich, Markus; Ongania, Karl-Hans; Kräutler, Bernhard (2007). "Colorless Tetrapyrrolic Chlorophyll Catabolites Found in Ripening Fruit Are Effective Antioxidants". Angewandte Chemie. 46 (45): 8699–8702. doi:10.1002/anie.200703587. PMC 2912502 . PMID 17943948.
21. Gross, Jeana (1991). Pigments in vegetables: chlorophylls and carotenoids. Van Nostrand Reinhold, ISBN 0442006578.
22. Larkum, edited by Anthony W. D. Larkum, Susan E. Douglas & John A. Raven (2003). Photosynthesis in algae. London: Kluwer. ISBN 0-7923-6333-7.

23. Cate, Thomas; Perkins, T. D. (September 2003). "Joseph Pelletier and Joseph Caventou". Journal of Tree Physiology. 23 (15): 1077–1079. doi:10.1093/treephys/23.15.1077.

24. Gitelson A. A., Buschmann C., Lichtenthaler H. K. (1999) "The Chlorophyll Fluorescence Ratio F735/F700 as an Accurate Measure of Chlorophyll Content in Plants" Remote Sens. Enviro. 69:296-302 (1999)

25. Meskauskiene R; Nater M; Goslings D; Kessler F; op den Camp R; Apel K. (23 October 2001). "FLU: A negative regulator of chlorophyll biosynthesis in Arabidopsis thaliana". Proceedings of the National Academy of Sciences. 98 (22): 12826–12831. Bibcode:2001PNAS...9812826M. doi:10.1073/pnas.221252798. JSTOR 3056990. PMC 60138 . PMID 11606728.

26. Duble, Richard L. "Iron Chlorosis in Turfgrass". Texas A&M University. Retrieved 2010-07-17.

27. Karageorgou, P.; Manetas, Y. (2006). "The importance of being red when young: Anthocyanins and the protection of young leaves of Quercus coccifera from insect herbivory and excess light". Tree Physiology. 26 (5): 613–21. doi:10.1093/treephys/26.5.613. PMID 16452075.

28. Chlorophyll : Global Maps. Earthobservatory.nasa.gov. Retrieved on 2014-02-02.

29. Adams, Jad (2004). Hideous absinthe : a history of the devil in a bottle. United Kingdom: I.B.Tauris, 2004. p. 22. ISBN 1860649203.

30. Lu, Congming; Zhang, Jianhua (July 1999). "Effects of Water Stress on Photosystem II Photochemistry and Its Thermostability in Wheat Plants" (PDF). Oxford Journals. Retrieved http://jxb.oxfordjournals.org/content/50/336/1199.full.pdf. Check date values in: |access-date= (help)

31. Zhu X-G., Govindjee, Baker N.R., deSturler E., Ort D.R., Long S.P. (2005) Chlorophyll a fluorescence induction kinetics in leaves predicted from a model describing each discrete step of excitation energy and electron transfer associated with Photosystem II Planta (2005) 223: 114–133 DOI 10.1007/s00425-005-0064-4

32. Zhu X-G., Govindjee, Baker N.R., deSturler E., Ort D.R., Long S.P. (2005) Chlorophyll a fluorescence induction kinetics in leaves predicted from a model describing each discrete step of excitation energy and electron transfer associated with Photosystem II Planta (2005) 223: 114–133 DOI 10.1007/s00425-005-0064-4

33. Zhu X-G., Govindjee, Baker N.R., deSturler E., Ort D.R., Long S.P. (2005) Chlorophyll a fluorescence induction kinetics in leaves predicted from a model describing each discrete step of excitation energy and electron transfer associated with Photosystem II Planta (2005) 223: 114–133 DOI 10.1007/s00425-005-0064-4

34. "Chlorophyll fluorescence—a practical guide" (PDF). Jxb.oxfordjournals.org. 2000-04-01. Retrieved 2011-03-28.

35. "Effects of Boron and Salinity on Red Raspberry in Vitro - International Journal of Fruit Science". Informaworld.com. 2008-12-03. Retrieved 2011-03-28.

36. Kitajima M, Butler WL (1975). "Quenching of chlorophyll fluorescence and primary photochemistry in chloroplasts by dibromothymoquinone.". Biochim

Biophys Acta. 376: 105–115. doi:10.1016/0005-2728(75)90209-1.

37. Genty B, Briantais J-M, Baker NR (1989). "The relationship between the quantum yield of photosynthetic electron transport and quenching of chlorophyll fluorescence.". Biochem Biophys Acta. 990: 87–92. doi:10.1016/s0304-4165(89)80016-9.
38. Schreiber U, Schliwa U, Bilger W. "Continuous recording of photochemical and non-photochemical chlorophyll fluorescence quenching with a new type of modulation fluorometer.". Photosynth Res. 10: 51–62. doi:10.1007/bf00024185.
39. Sobrado (2008). "Leaf characteristics and diurnal variation of chlorophyll fluorescence in leaves of the 'bana' vegetation of the amazon region" (PDF).
40. "Plant Stress Biology". Personalpages.manchester.ac.uk. Retrieved 2011-03-28.
41. Favaretto; et al. (2011). "Differential responses of antioxidant enzymes in pioneer and late-successional tropical tree species grown under sun and shade conditions". Environmental and Experimental Botany. 70: 20–28. doi:10.1016/j.envexpbot.2010.06.003.
42. Lu, Congming, and Jianhua Zhang. "Effects of Water Stress on Photosystem II Photochemistry and Its Thermostability in Wheat Plants." Journal of Experimental Botany No. 336, Pp. 1199–1206, July 1999 Effects of Water Stress on Photosystem II Photochemistry and Its Thermostability in Wheat Plants 50.336 (1999): 1199-206. Oxfordjournals.org. Oxford Journals. Web. 8 Mar. 2016.
43. A. Cartelat; Z.G. Cerovic; Y. Goulas; S. Meyer; C. Lelarge; J.-L. Prioul; A. Barbottin; M.-H. Jeuffroy; P. Gate; G. Agati; I. Moya (2005). "Optically assessed contents of leaf polyphenolics and chlorophyll as indicators of nitrogen deficiency in wheat (Triticum aestivum L.)". Field Crops Research. Field Crops Research Volume 91, Issue 1, pages 35-49. 91: 35–49. doi:10.1016/j.fcr.2004.05.002.
44. Gitelson, Anatoly A; Buschmann, Claus; Lichtenthaler, Hartmut K (1999). "The Chlorophyll Fluorescence Ratio F735/F700 as an Accurate Measure of the Chlorophyll Content in Plants". Remote Sensing of Environment. 69 (3): 296–302. doi:10.1016/S0034-4257(99)00023-1.
45. Schreiber U, Bilger W, Schliwa U (1986). "Continuous recording of photochemical and non-photochemical chlorophyll fluorescence quenching with a new type of modulation fluorometer". Photosynth. Res. 10: 51–62. doi:10.1007/bf00024185.
46. Schreiber, Ulrich (1986). "Detection of rapid induction kinetics with a new type of high-frequency modulated chlorophyll fluorometer". Photosynth. Res. 9: 261–272. doi:10.1007/bf00029749.
47. Baker N. R., Oxborough K., (2004) Chlorophyll fluorescence as a probe of photosynthetic productivity. From Chapter 3, "Chlorophyll a Fluorescence a Signature of Photosynthesis", edited by George Papaqeorgiou and Govindjee, published by Springer 2004, PO Box 17, 3300 AA Dordrecht, The Netherlands, pages 66-79
48. Genty B., Briantais J. M. & Baker N. R. (1989) The relationship between the quantum yield of photosynthetic electron transport and quenching of chlorophyll fluorescence, Biochimica et Biophysica Acta 990, 87-92

49. Haldimann P, & Feller U. (2004) Inhibition of photosynthesis by high temperature in oak (Quercus pubescens L.) leaves grown under natural conditions closely correlates with a reversible heat dependent reduction of the activation state of ribulose-1,5-bisphosphate carboxylase/oxygenase.

50. Kramer D. M., Johnson G., Kiirats O., Edwards G. (2004) New fluorescence parameters for determination of QA redox state and excitation energy fluxes. Photosynthesis Research 79: 209-218

51. van Kooten O, Snel J (1990) The use of chlorophyll fluorescence nomenclature in plant stress physiology. Photosynth Res 25: 147–150

52. Klughammer C., and Schreiber U. (2008) PAM Application notes 2008 1:27 -35

53. Muller P., Xiao-Ping L., Niyogi K. (2001) Non-Photochemical Quenching. A Response to Excess Light Energy. Plant Physiology 125, 1558-1556

54. Ruban A.V., Murchie E.H. (2012) Assessing the photoprotective effectiveness of non-photochemical chlorophyll fluorescence quenching: A new approach Biochimica et Biophysica Acta 1817 (2012) 977–982

55. Ruban A.V., Johnson M.P., (2009) Dynamics of higher plant photosystem cross-section associated with state transitions. Photosynthesis Research 2009 99:173-183

56. Cazzaniga S, Osto L.D., Kong S-G., Wada M., Bassi R., (2013) "Interaction between avoidance of photon absorption, excess energy dissipation and zeaxanthin synthesis against photooxidative stress in Arabidopsis", The Plant Journal, Volume 76, Issue 4, pages 568–579, November 2013 DOI: 10.1111/tpj.12314

57. Lichtenthaler H. K., Babani F. (2004) Light Adaption and Senescence of the Photosynthetic Apparatus. Changes in Pigment Composition, Chlorophyll Fluorescence Parameters and Photosynthetic Activity. From Chapter 28, "Chlorophyll a Fluorescence a Signature of Photosynthesis", edited by George Papaqeorgiou and Govindjee, published by Springer 2004,PO Box 17, 3300 AA Dordrecht, The Netherlands, page 716

58. Handy PEA: Continuous Excitation Plant Efficiency Analyser (PDF). Norfolk: Hansatech Instruments. 2012. p. 2.

59. Lavrov; et al. (2012). "Water stress assessment of cork oak leaves and maritime pine needles based on LIF spectra". Optics and Spectroscopy. Springer. 112 (2): 271–279. doi:10.1134/S0030400X12020166.

60. Silvestre et al. Contribution of pPLAIIα to drought tolerance using genetically modified arabidopsis plants: II. Effects on photosynthetic metabolism. Int. Meeting Prog. Plant Symposium of the SEB: Oxidative stress and cell death in plants: mechanisms and implications, Florence, Italy, 26–28 June 2013, p. 5

61. Utkin; et al. (2013). "Compact low-cost detector for in vivo assessment of microphytobenthos using laser induced fluorescence". Optics and Spectroscopy. Springer. 114 (3): 471–477. doi:10.1134/S0030400X13030259.

62. Vieira; et al. (2011). "Effects of intertidal microphytobenthos migration on biomass determination via laser-induced fluorescence". Marine Ecology Progress Series. 432: 45–52. doi:10.3354/meps09157.

CHAPTER 3

The Role of Chloroplast

Chapter Contents

- **Introduction**
- **Discovery**
- **Chloroplast DNA**
- **Function and Chemistry**
- **Chloroplast DNA**

A chloroplast is a type of structure, called an organelle that is found in plants and is where photosynthesis occurs. Normally present in plant leaves, chloroplasts contain all of the components that allow the plant to convert sunlight into usable energy. The main components of chloroplasts are the membranes, chlorophyll and other pigments, grana and stroma.

Chloroplasts are one of the most important components of a plant because the entire photosynthetic process takes place in them. Each cell in a plant leaf can have 50 of these organelles. Chloroplasts appear only in eukaryokic organisms, which are primarily non-animal.

INTRODUCTION

Chloroplasts are organelles, specialized subunits, in plant and algal cells. Their discovery inside plant cells is usually credited to Julius von Sachs (1832–1897), an influential botanist and author of standard botanical textbooks – sometimes called "The Father of Plant Physiology".

Chloroplasts' main role is to conduct photosynthesis, where the photosynthetic pigment chlorophyll captures the energy from sunlight and converts it and stores it in the energy-storage molecules ATP and NADPH while freeing oxygen from water. They then use the ATP and NADPH to make organic molecules from carbon dioxide in a process known as the Calvin cycle. Chloroplasts carry out a number of other functions, including fatty acid synthesis, much amino acid synthesis, and the immune response in plants. The number of chloroplasts per cell varies from one, in unicellular algae, up to 100 in plants like Arabidopsis and wheat.

A chloroplast is a type of organelle known as a plastid, characterized by its high concentration of chlorophyll. Other plastid types, such as the leucoplast and the chromoplast, contain little chlorophyll and do not carry out photosynthesis.

Chloroplasts are highly dynamic—they circulate and are moved around within plant cells, and occasionally pinch in two to reproduce. Their behavior is strongly influenced by environmental factors like light color and intensity. Chloroplasts, like mitochondria, contain their own DNA, which is thought to be inherited from their ancestor—a photosynthetic cyanobacterium that was engulfed by an early eukaryotic cell. Chloroplasts cannot be made by the plant cell and must be inherited by each daughter cell during cell division.

With one exception (the amoeboid *Paulinella chromatophora*), all chloroplasts can probably be traced back to a single endosymbiotic event, when a cyanobacterium was engulfed by the eukaryote. Despite this, chloroplasts can be found in an extremely wide set of organisms, some not even directly related to each other—a consequence of many secondary and even tertiary endosymbiotic events.

The word *chloroplast* (Greek: χλωροπλάστης) is derived from the Greek words *chloros* , which means green, and *plastes* , which means "the one who forms".

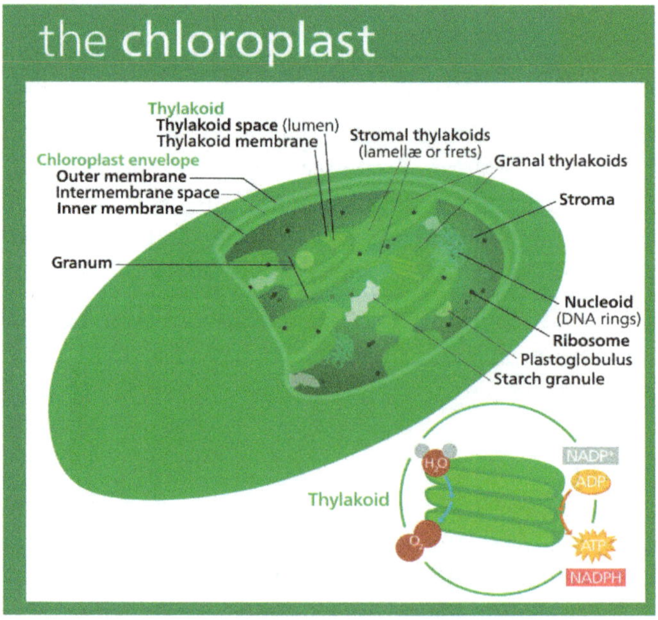

Structure of a typical higher-plant chloroplast

DISCOVERY

The first definitive description of a chloroplast (*Chlorophyllkörnen*, "grain of chlorophyll") was given by Hugo von Mohl in 1s837 as discrete bodies within the green plant cell. In 1883, A. F. W. Schimper would name these bodies as "chloroplastids" (*Chloroplastiden*). In 1884, Eduard Strasburger adopted the term "chloroplasts" (*Chloroplasten*).

Chloroplast Lineages and Evolution

Chloroplasts are one of many types of organelles in the plant cell. They are considered to have originated from cyanobacteria through endosymbiosis—when a eukaryotic cell engulfed a photosynthesizing cyanobacterium that became a permanent resident in the cell. Mitochondria are thought to have come from a similar event, where an aerobic prokaryote was engulfed. This origin of chloroplasts was first suggested by the Russian biologist Konstantin Mereschkowski in 1905after Andreas Schimper observed in 1883 that chloroplasts closely resemble cyanobacteria. Chloroplasts are only found in plants, algae, and the amoeboid*Paulinella chromatophora*.

Cyanobacterial Ancestor

Cyanobacteria are considered the ancestors of chloroplasts. They are sometimes called blue-green algae even though they are prokaryotes. They are a diverse phylum of bacteria capable of carrying out photosynthesis, and are gram-negative, meaning that they have two cell membranes. Cyanobacteria also contain a peptidoglycan cell wall, which is thicker than in other gram-negative bacteria, and which is located between their two cell membranes. Like chloroplasts, they have thylakoids within. On the thylakoid membranes are photosynthetic pigments, including chlorophyll *a*. Phycobilins are also common cyanobacterial pigments, usually organized into hemispherical phycobilisomes attached to the outside of the thylakoid membranes (phycobilins are not shared with all chloroplasts though).

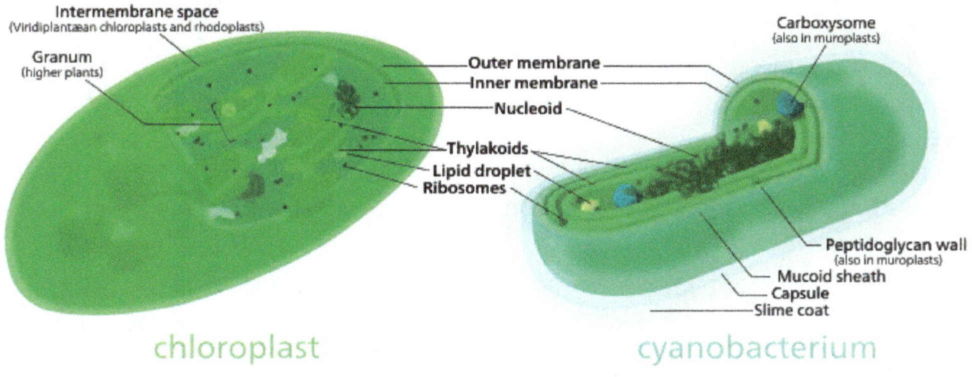

Both chloroplasts and cyanobacteria have a double membrane, DNA, ribosomes, and thylakoids. Both the chloroplast and cyanobacterium depicted are idealized versions (the chloroplast is that of a higher plant)—a lot of diversity exists among chloroplasts and cyanobacteria.

Primary Endosymbiosis

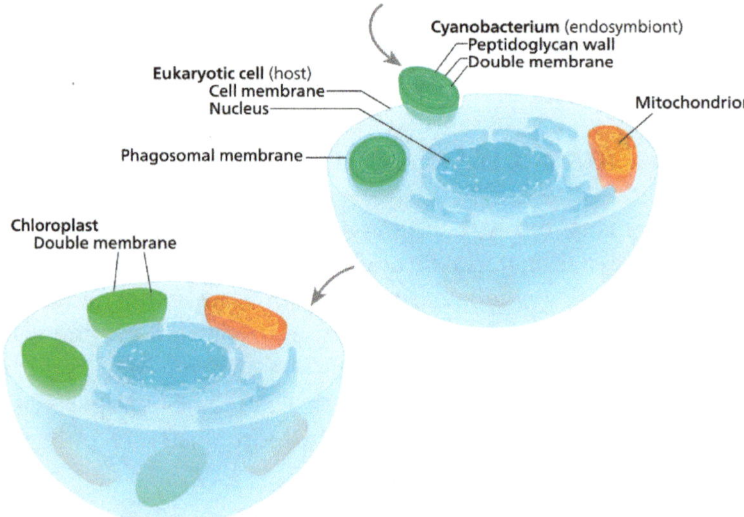

Primary Endosymbiosis

A eukaryote with mitochondria engulfed a cyanobacterium in an event of serial primary endosymbiosis, creating a lineage of cells with both organelles. It is important to note that the cyanobacterial endosymbiont already had a double membrane—the phagosomal vacuole-derived membrane was lost.

Somewhere around a billion years ago, a free-living cyanobacterium entered an early eukaryotic cell, either as food or as an internal parasite,but managed to escape the phagocytic vacuole it was contained in. The two innermost lipid-bilayer membranes that surround all chloroplasts correspond to the outer and inner membranes of the ancestral cyanobacterium's gram negative cell wall, and not the phagosomal membrane from the host, which was probably lost. The new cellular resident quickly became an advantage, providing food for the eukaryotic host, which allowed it to live within it. Over time, the cyanobacterium was assimilated, and many of its genes were lost or transferred to the nucleus of the host. Some of its proteins were then synthesized in the cytoplasm of the host cell, and imported back into the chloroplast (formerly the cyanobacterium). Separately, somewhere around 100 million years ago, it happened again and led to the amoeboid *Paulinella chromatophora*.

This event is called *endosymbiosis*, or "cell living inside another cell". The cell living inside the other cell is called the *endosymbiont*; the endosymbiont is found inside the *host cell*.

Chloroplasts are believed to have arisen after mitochondria, since all eukaryotes contain mitochondria, but not all have chloroplasts. This is called *serial endosymbiosis*—an early eukaryote engulfing the mitochondrion ancestor, and some descendants of it then engulfing the chloroplast ancestor, creating a cell with both chloroplasts and mitochondria.

Whether or not chloroplasts came from a single endosymbiotic event, or many independent engulfments across various eukaryotic lineages, has been long debated. It is now generally held that most organisms with chloroplasts either share a single ancestor or obtained their chloroplast from organisms that share a common ancestor that took in a cyanobacterium 600–1600 million years ago. The exception is the amoeboid *Paulinella chromatophora*, which descends from an ancestor that took in a cyanobacterium 90–140 million years ago.

These chloroplasts, which can be traced back directly to a cyanobacterial ancestor, are known as *primary plastids* (*«plastid»* in this context means almost the same thing as chloroplast). All primary chloroplasts belong to one of four chloroplast lineages—the glaucophyte chloroplast lineage, the amoeboid *Paulinella chromatophora* lineage, the rhodophyte (red algal) chloroplast lineage, or the chloroplastidan (green) chloroplast lineage. The rhodophyte and chloroplastidan lineages are the largest, with chloroplastidan (green) being the one that contains the land plants.

Chloroplast lineages A primary endosymbiosis event gave rise to four main lineages of chloroplasts in the glaucophytes, Paulinella, chlorophyta, and rhodophyta. Some of these algae were subsequently engulfed by other algae, becoming secondary (or tertiary) endosymbionts. [a] **The apicomplexans (malaria parasites), contain a red algal endosymbiont with a non-photosynthetic chloroplast.**[b] **2–3 chloroplast membranes**[a c] **2–4 chloroplast membranes**

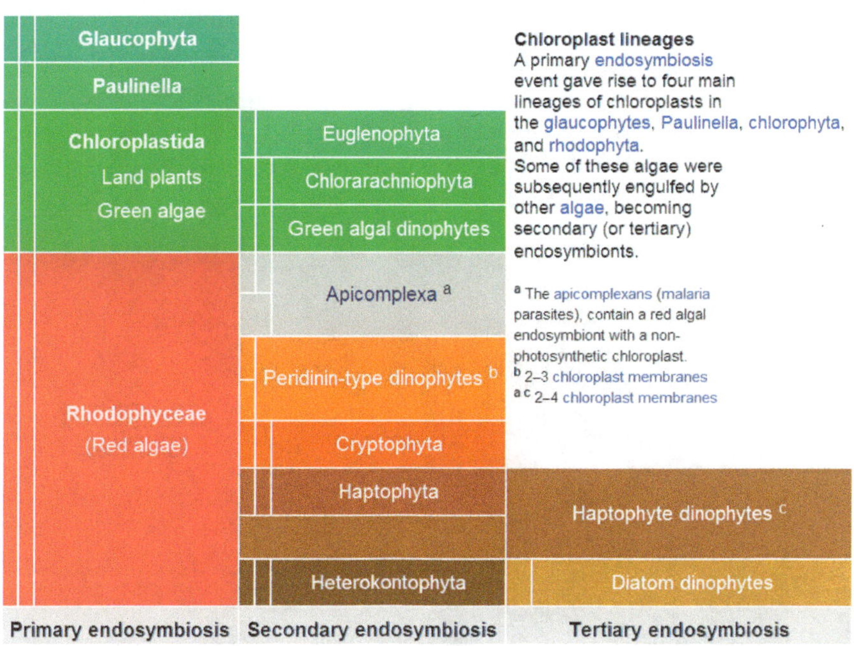

Glaucophyta

The alga *Cyanophora*, a glaucophyte, is thought to be one of the first organisms to contain a chloroplast. The glaucophyte chloroplast group is the smallest of the three primary chloroplast lineages, being found in only 13 species, and is thought to be the one that branched off the earliest. Glaucophytes have chloroplasts that retain a peptidoglycan wall between their double membranes, like their cyanobacterial parent. For this reason, glaucophyte chloroplasts are also known as *muroplasts*. Glaucophyte chloroplasts also contain concentric unstacked thylakoids, which surround a carboxysome - an icosahedral structure that glaucophyte chloroplasts and cyanobacteria keep their carbon fixation enzyme rubisco in. The starch that they synthesize collects outside the chloroplast. Like cyanobacteria, glaucophyte chloroplast thylakoids are studded with light collecting structures called phycobilisomes. For these reasons, glaucophyte chloroplasts are considered a primitive intermediate between cyanobacteria and the more evolved chloroplasts in red algae and plants.

Diversity of red algae Clockwise from top left: *Bornetia secundiflora*, *Peyssonnelia squamaria*, *Cyanidium*, *Laurencia*, *Callophyllis laciniata*. Red algal chloroplasts are characterized by phycobilin pigments which often give them their reddish color.

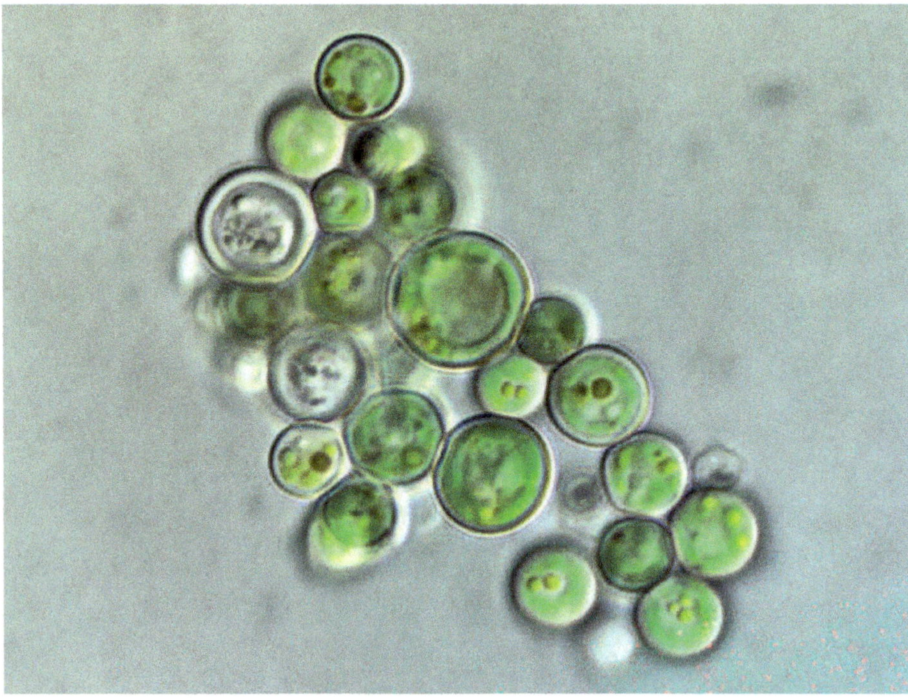

Rhodophyceae (Red Algae)

The rhodophyte, or red algal chloroplast group is another large and diverse chloroplast lineage. Rhodophyte chloroplasts are also called *rhodoplasts*, literally "red chloroplasts".

Rhodoplasts have a double membrane with an intermembrane space and phycobilin pigments organized into phycobilisomes on the thylakoid membranes, preventing their thylakoids from stacking. Some contain pyrenoids. Rhodoplasts have chlorophyll *a* and phycobilins for photosynthetic pigments; the phycobilin phycoerytherin is responsible for giving many red algae their distinctive red color. However, since they also contain the blue-green chlorophyll *a* and other pigments, many are reddish to purple from the combination. The red phycoerytherin pigment is an adaptation to help red algae catch more sunlight in deep water—as such, some red algae that live in shallow water have less phycoerytherin in their rhodoplasts, and can appear more greenish.Rhodoplasts synthesize a form of starch called floridean starch, which collects into granules outside the rhodoplast, in the cytoplasm of the red alga.

Chloroplastida (Green Algae and Plants)

Diversity of green algae Clockwise from top left: *Scenedesmus*, *Micrasterias*, *Hydrodictyon*, *Stigeoclonium*, *Volvox*. Green algal chloroplasts are characterized by their pigments chlorophyll *a* and chlorophyll *b* which give them their green color.

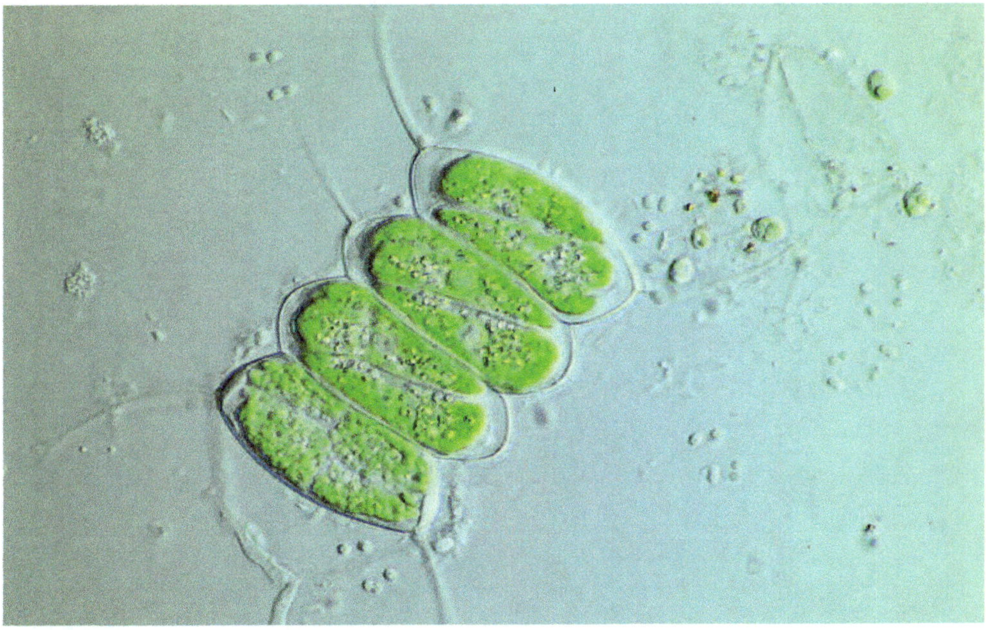

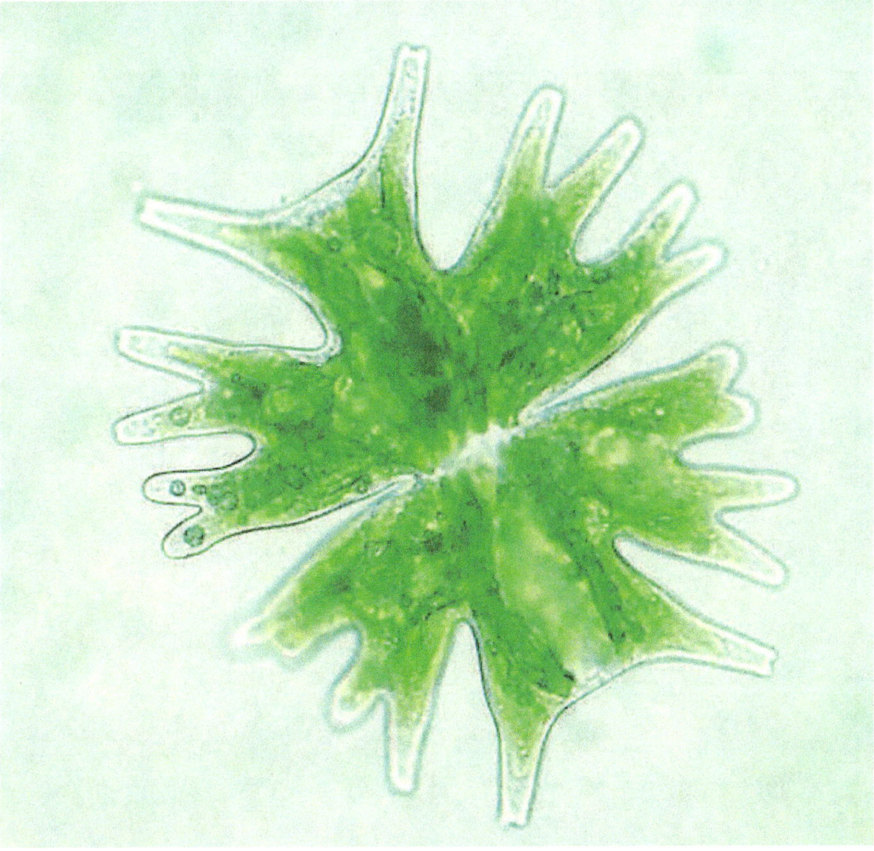

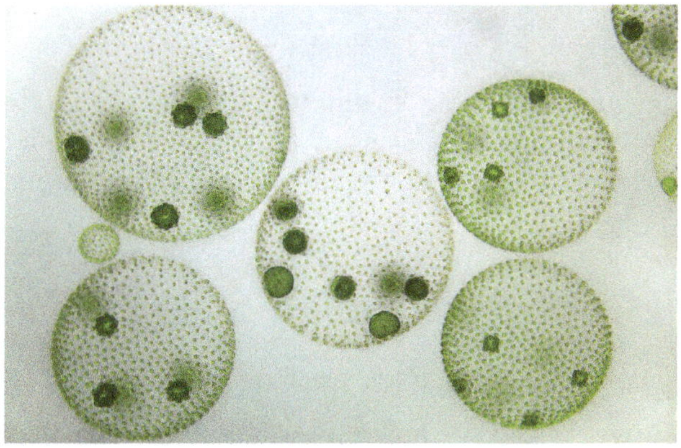

The chloroplastidan chloroplasts, or green chloroplasts, are another large, highly diverse primary chloroplast lineage. Their host organisms are commonly known as the green algae and land plants. They differ from glaucophyte and red algal chloroplasts in that they have lost their phycobilisomes, and contain chlorophyll *b* instead. Most green chloroplasts are (obviously) green, though some aren't, like some forms of *Hæmatococcus pluvialis*, due to accessory pigments that override the chlorophylls' green colors. Chloroplastidan chloroplasts have lost the peptidoglycan wall between their double membrane, leaving an intermembrane space. Some plants seem to have kept the genes for the synthesis of the peptidoglycan layer, though they've been repurposed for use in chloroplast division instead.

Most of the chloroplasts depicted in this article are green chloroplasts.

Green algae and plants keep their starch *inside* their chloroplasts, and in plants and some algae, the chloroplast thylakoids are arranged in grana stacks. Some green algal chloroplasts contain a structure called a pyrenoid, which is functionally similar to the glaucophyte carboxysome in that it is where rubisco and CO_2 are concentrated in the chloroplast.

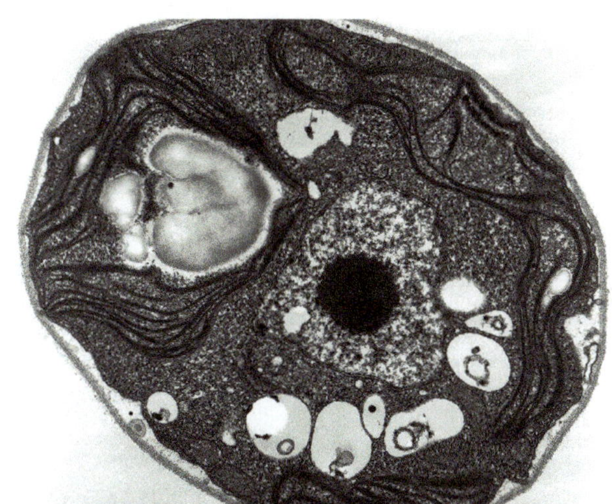

Transmission electron micrograph of *Chlamydomonas reinhardtii*, a green alga that contains a pyrenoid surrounded by starch.

Helicosporidium

Helicosporidium is a genus of nonphotosynthetic parasitic green algae that is thought to contain a vestigial chloroplast. Genes from a chloroplast and nuclear genes indicating the presence of a chloroplast have been found in Helicosporidiumeven if nobody's seen the chloroplast itself.

Secondary and Tertiary Endosymbiosis

Many other organisms obtained chloroplasts from the primary chloroplast lineages through secondary endosymbiosis—engulfing a red or green alga that contained a chloroplast. These chloroplasts are known as secondary plastids.

While primary chloroplasts have a double membrane from their cyanobacterial ancestor, secondary chloroplasts have additional membranes outside of the original two, as a result of the secondary endosymbiotic event, when a nonphotosynthetic eukaryote engulfed a chloroplast-containing alga but failed to digest it—much like the cyanobacterium at the beginning of this story. The engulfed alga was broken down, leaving only its chloroplast, and sometimes its cell membrane and nucleus, forming a chloroplast with three or four membranes—the two cyanobacterial membranes, sometimes the eaten alga's cell membrane, and the phagosomal vacuole from the host's cell membrane.

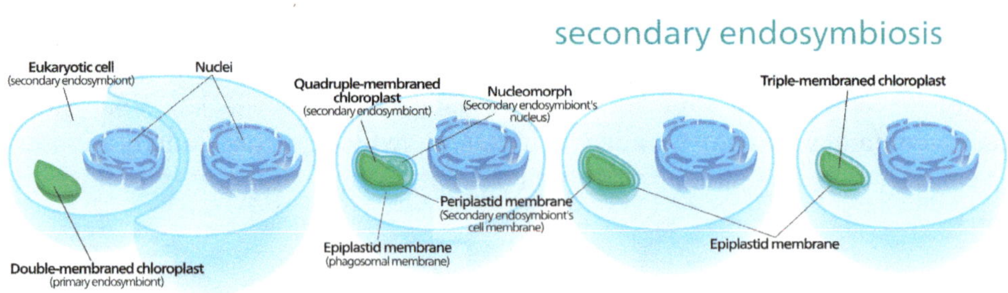

Secondary endosymbiosis consisted of a eukaryotic alga being engulfed by another eukaryote, forming a chloroplast with three or four membranes.

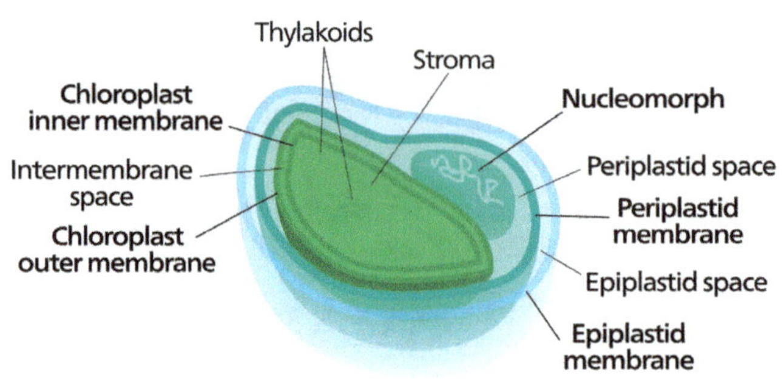

Diagram of a four membraned chloroplast containing a nucleomorph.

The genes in the phagocytosed eukaryote's nucleus are often transferred to the secondary host's nucleus.Cryptomonads and chlorarachniophytes retain the phagocytosed eukaryote's nucleus, an object called a nucleomorph, located between the second and third membranes of the chloroplast.

All secondary chloroplasts come from green and red algae—no secondary chloroplasts from glaucophytes have been observed, probably because glaucophytes are relatively rare in nature, making them less likely to have been taken up by another eukaryote.

Green algal Derived Chloroplasts

Green algae have been taken up by the euglenids, chlorarachniophytes, a lineage of dinoflagellates, and possibly the ancestor of the chromalveolates in three or four separate engulfments. Many green algal derived chloroplasts contain pyrenoids, but unlike chloroplasts in their green algal ancestors, starch collects in granules outside the chloroplast.

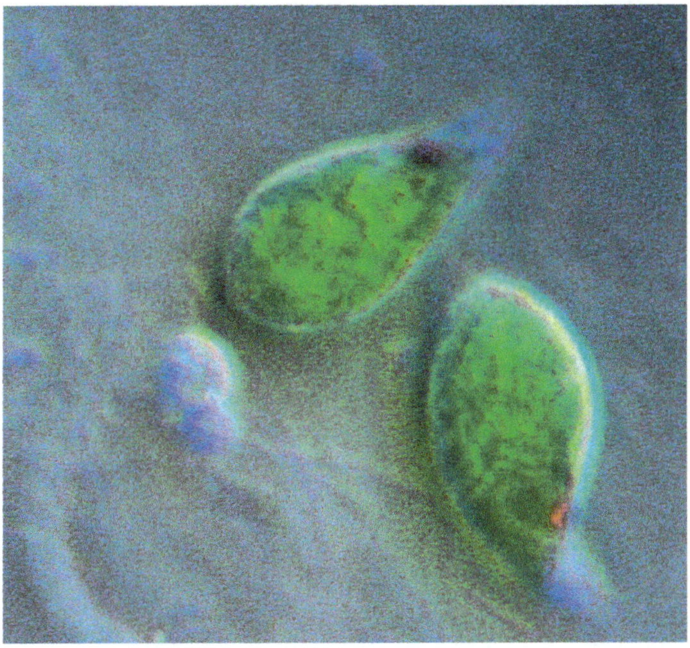

Euglena, a euglenophyte, contains secondary chloroplasts from green algae.

Euglenophytes

Euglenophytes are a group of common flagellated protists that contain chloroplasts derived from a green alga.Euglenophyte chloroplasts have three membranes—it is thought that the membrane of the primary endosymbiont was lost, leaving the cyanobacterial membranes, and the secondary host's phagosomal membrane. Euglenophyte chloroplasts have a pyrenoid and thylakoids stacked in groups of three. Starch is stored in the form of paramylon, which is contained in membrane-bound granules in the cytoplasm of the euglenophyte.

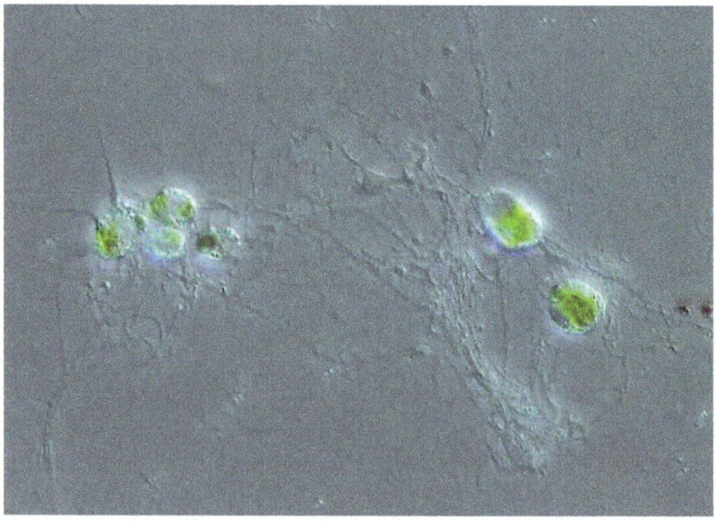

Chlorarachnion reptans is a chlorarachniophyte. Chlorarachniophytes replaced their original red algal endosymbiont with a green alga.

Chlorarachniophytes

Chlorarachniophytes are a rare group of organisms that also contain chloroplasts derived from green algae, though their story is more complicated than that of the euglenophytes. The ancestor of chlorarachniophytes is thought to have been a chromalveolate, a eukaryote with a *red* algal derived chloroplast. It is then thought to have lost its first red algal chloroplast, and later engulfed a green alga, giving it its second, green algal derived chloroplast.

Chlorarachniophyte chloroplasts are bounded by four membranes, except near the cell membrane, where the chloroplast membranes fuse into a double membrane. Their thylakoids are arranged in loose stacks of three. Chlorarachniophytes have a form of starch called chrysolaminarin, which they store in the cytoplasm, often collected around the chloroplast pyrenoid, which bulges into the cytoplasm.

Chlorarachniophyte chloroplasts are notable because the green alga they are derived from has not been completely broken down—its nucleus still persists as a nucleomorph found between the second and third chloroplast membranes—the periplastid space, which corresponds to the green alga›s cytoplasm.

Early Chromalveolates

Recent research has suggested that the ancestor of the chromalveolates acquired a green algal prasinophyte endosymbiont. The green algal derived chloroplast was lost and replaced with a red algal derived chloroplast, but not before contributing some of its genes to the early chromalveolate's nucleus. The presence of both green algal and red algal genes in chromalveolates probably helps them thrive under fluctuating light conditions.

Red algal Derived Chloroplasts (Chromalveolate Chloroplasts)

Like green algae, red algae have also been taken up in secondary endosymbiosis, though it is thought that all red algal derived chloroplasts are descended from a single red alga that was engulfed by an early chromalveolate, giving rise to the chromalveolates, some of which, like the ciliates, subsequently lost the chloroplast. This is still debated though.

Pyrenoids and stacked thylakoids are common in chromalveolate chloroplasts, and the outermost membrane of many are continuous with the rough endoplasmic reticulum and studded with ribosomes. They have lost their phycobilisomes and exchanged them for chlorophyll *c*, which isn't found in primary red algal chloroplasts themselves.

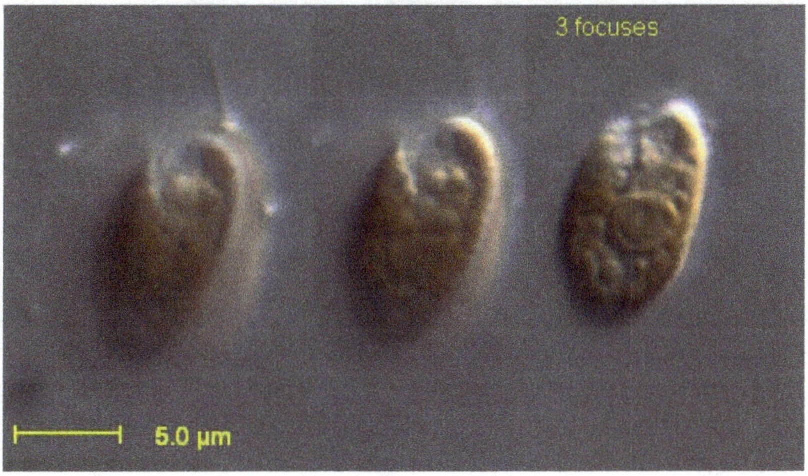

Rhodomonas salina is a cryptophyte.

Cryptophytes

Cryptophytes, or cryptomonads are a group of algae that contain a red-algal derived chloroplast. Cryptophyte chloroplasts contain a nucleomorph that superficially resembles that of the chlorarachniophytes. Cryptophyte chloroplasts have four membranes, the outermost of which is continuous with the rough endoplasmic reticulum. They synthesize ordinary starch, which is stored in granules found in the periplastid space—outside the original double membrane, in the place that corresponds to the red alga's cytoplasm. Inside cryptophyte chloroplasts is a pyrenoid and thylakoids in stacks of two.

Their chloroplasts do not have phycobilisomes, but they do have phycobilin pigments which they keep in their thylakoid space, rather than anchored on the outside of their thylakoid membranes.

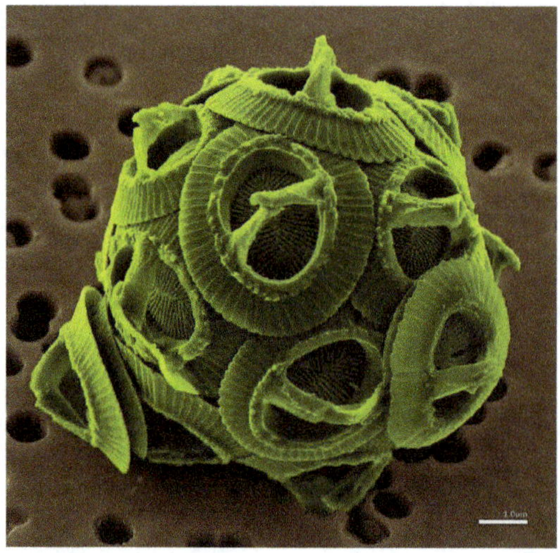

Scanning electron micrograph of *Gephyrocapsa oceanica*, a haptophyte.

Haptophytes

Haptophytes are similar and closely related to cryptophytes, and are thought to be the first chromalveolates to branch off.Their chloroplasts lack a nucleomorph, their thylakoids are in stacks of three, and they synthesize chrysolaminarin sugar, which they store completely outside of the chloroplast, in the cytoplasm of the haptophyte.

Heterokontophytes (Stramenopiles)

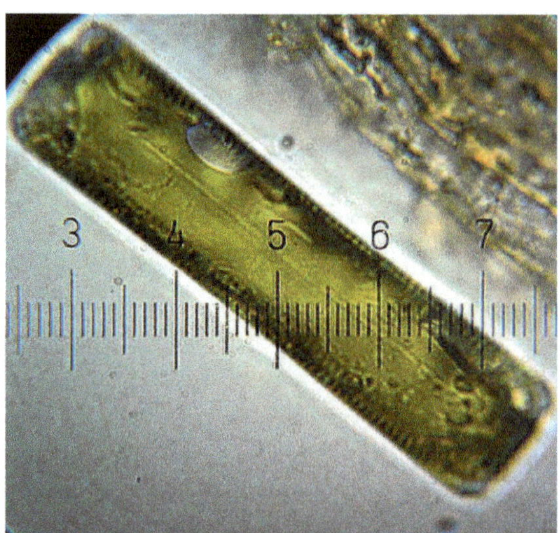

The photosynthetic pigments present in their chloroplasts give diatoms a greenish-brown color.

The heterokontophytes, also known as the stramenopiles, are a very large and diverse group of algae that also contain red algal derived chloroplasts. Heterokonts include the diatoms and the brown algae, golden algae, and yellow-green algae.

Heterokont chloroplasts are very similar to haptophyte chloroplasts, containing a pyrenoid, triplet thylakoids, and with some exceptions, having an epiplastid membrane connected to the endoplasmic reticulum. Like haptophytes, heterokontophytes store sugar in chrysolaminarin granules in the cytoplasm. Heterokontophyte chloroplasts contain chlorophyll *a* and with a few exceptions chlorophyll *c*, but also have carotenoids which give them their many colors.

Apicomplexans

Apicomplexans are another group of chromalveolates. Like the helicosproidia, they're parasitic, and have a nonphotosynthetic chloroplast. They were once thought to be related to the helicosproidia, but it is now known that the helicosproida are green algae rather than chromalveolates. The apicomplexans include *Plasmodium*, the malaria parasite. Many apicomplexans keep a vestigial red algal derived chloroplast called an apicoplast, which they inherited from their ancestors. Other apicomplexans like *Cryptosporidium* have lost the chloroplast completely. Apicomplexans store their energy in amylopectin starch granules that are located in their cytoplasm, even though they are nonphotosynthetic.

Apicoplasts have lost all photosynthetic function, and contain no photosynthetic pigments or true thylakoids. They are bounded by four membranes, but the membranes are not connected to the endoplasmic reticulum. The fact that apicomplexans still keep their nonphotosynthetic chloroplast around demonstrates how the chloroplast carries out important functions other than photosynthesis. Plant chloroplasts provide plant cells with many important things besides sugar, and apicoplasts are no different—they synthesize fatty acids, isopentenyl pyrophosphate, iron-sulfur clusters, and carry out part of the heme pathway. This makes the apicoplast an attractive target for drugs to cure apicomplexan-related diseases. The most important apicoplast function is isopentenyl pyrophosphate synthesis—in fact, apicomplexans die when something interferes with this apicoplast function, and when apicomplexans are grown in an isopentenyl pyrophosphate-rich medium, they dump the organelle.

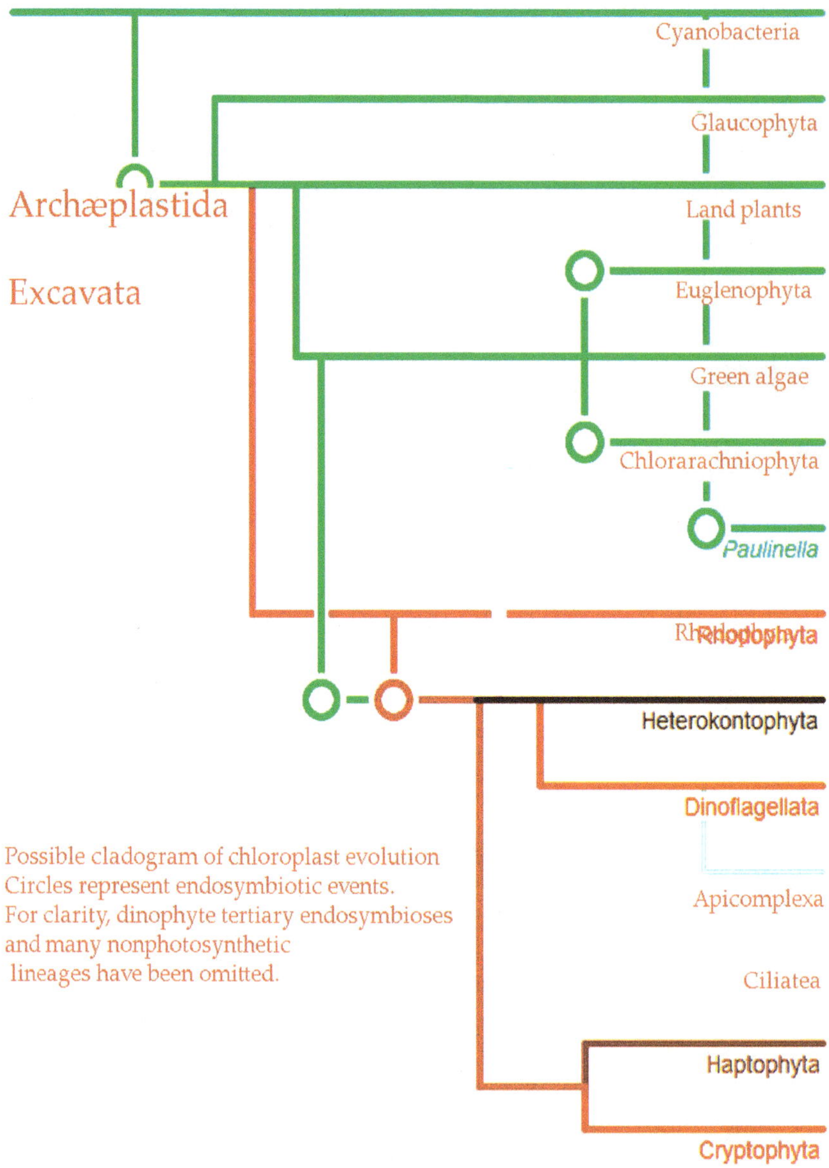

Possible cladogram of chloroplast evolution. Circles represent endosymbiotic events. For clarity, dinophyte tertiary endosymbioses and many nonphotosynthetic lineages have been omitted.

Dinophytes

The dinoflagellates are yet another very large and diverse group of protists, around half of which are (at least partially) photosynthetic.

Most dinophyte chloroplasts are secondary red algal derived chloroplasts, like other chromalveolate chloroplasts. Many other dinophytes have lost the chloroplast (becoming the nonphotosynthetic kind of dinoflagellate), or replaced it though *tertiary* endosymbiosis—the engulfment of another chromalveolate containing a red algal derived chloroplast. Others replaced their original chloroplast with a green algal derived one.

Most dinophyte chloroplasts contain at least the photosynthetic pigments chlorophyll *a*, chlorophyll c_2, *beta*-carotene, and at least one dinophyte-unique xanthophyll (peridinin, dinoxanthin, or diadinoxanthin), giving many a golden-brown color. All dinophytes store starch in their cytoplasm, and most have chloroplasts with thylakoids arranged in stacks of three.

Peridinin-Containing Dinophyte Chloroplast

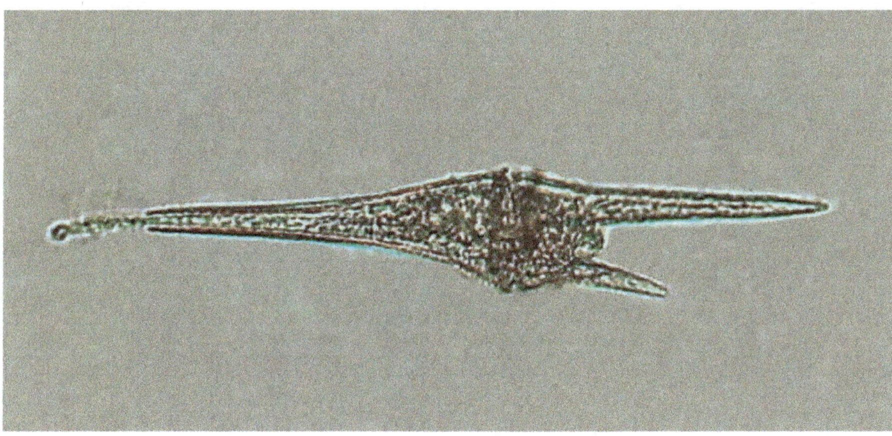

Ceratium furca, a peridinin-containing dinophyte

The most common dinophyte chloroplast is the peridinin-type chloroplast, characterized by the carotenoid pigment peridinin in their chloroplasts, along with chlorophyll *a* and chlorophyll c_2. Peridinin is not found in any other group of chloroplasts. The peridinin chloroplast is bounded by three membranes (occasionally two), having lost the red algal endosymbiont's original cell membrane. The outermost membrane is not connected to the endoplasmic reticulum. They contain a pyrenoid, and have triplet-stacked thylakoids. Starch is found outside the chloroplast. An important feature of these chloroplasts is that their chloroplast DNA is highly reduced and fragmented into many small circles. Most of the genome has migrated to the nucleus, and only critical photosynthesis-related genes remain in the chloroplast.

The peridinin chloroplast is thought to be the dinophytes' "original" chloroplast, which has been lost, reduced, replaced, or has company in several other dinophyte lineages.

Fucoxanthin-Containing Dinophyte Chloroplasts (Haptophyte Endosymbionts)

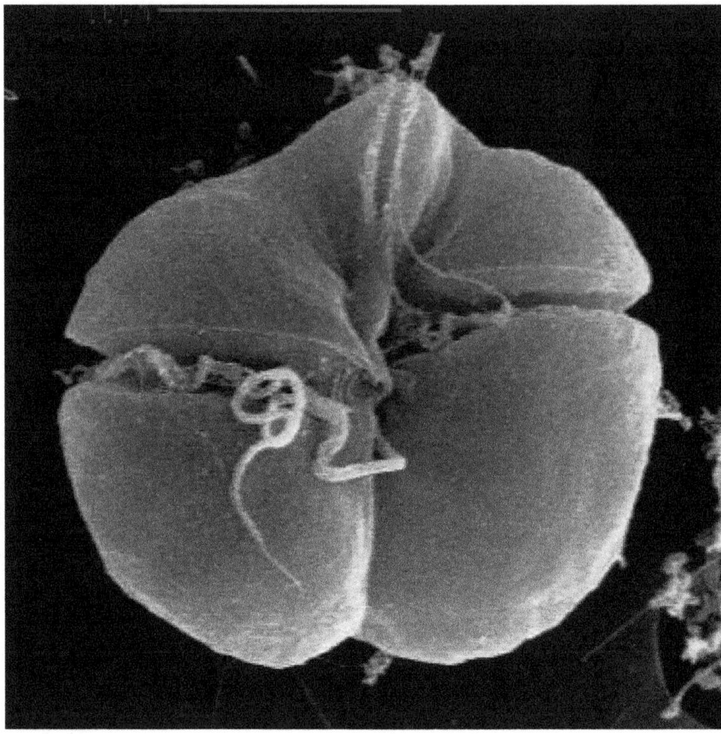

Karenia brevis is a fucoxanthin-containing dynophyte responsible for algal blooms called "red tides".

The fucoxanthin dinophyte lineages (including *Karlodinium* and *Karenia*) lost their original red algal derived chloroplast, and replaced it with a new chloroplast derived from a haptophyte endosymbiont. *Karlodinium* and *Karenia* probably took up different heterokontophytes. Because the haptophyte chloroplast has four membranes, tertiary endosymbiosis would be expected to create a six membraned chloroplast, adding the haptophyte's cell membrane and the dinophyte's phagosomal vacuole. However, the haptophyte was heavily reduced, stripped of a few membranes and its nucleus, leaving only its chloroplast (with its original double membrane), and possibly one or two additional membranes around it.

Fucoxanthin-containing chloroplasts are characterized by having the pigment fucoxanthin (actually 19′-hexanoyloxy-fucoxanthin and/or 19′-butanoyloxy-fucoxanthin) and no peridinin. Fucoxanthin is also found in haptophyte chloroplasts, providing evidence of ancestry.

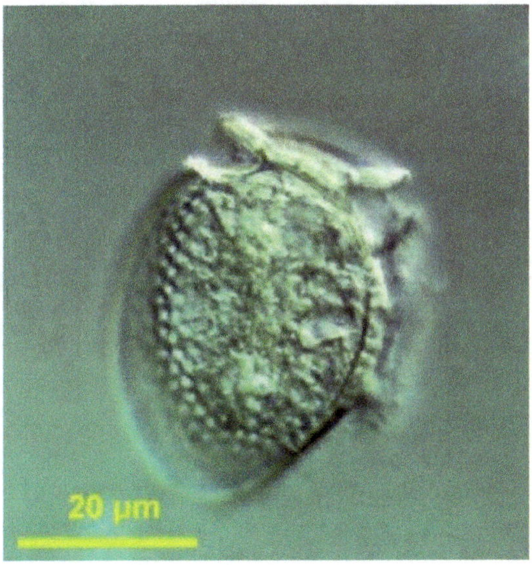

Dinophysis acuminata has chloroplasts taken from a cryptophyte.

Cryptophyte Derived Dinophyte Chloroplast

Members of the genus *Dinophysis* have a phycobilin-containing chloroplast taken from a cryptophyte. However, the cryptophyte is not an endosymbiont—only the chloroplast seems to have been taken, and the chloroplast has been stripped of its nucleomorph and outermost two membranes, leaving just a two-membraned chloroplast. Cryptophyte chloroplasts require their nucleomorph to maintain themselves, and *Dinophysis* species grown in cell culture alone cannot survive, so it is possible (but not confirmed) that the *Dinophysis* chloroplast is a kleptoplast—if so, *Dinophysis* chloroplasts wear out and *Dinophysis* species must continually engulf cryptophytes to obtain new chloroplasts to replace the old ones.

Diatom Derived Dinophyte Chloroplasts

Some dinophytes, like *Kryptoperidinium* and *Durinskia* have a diatom (heterokontophyte) derived chloroplast. These chloroplasts are bounded by up to *five* membranes, (depending on whether you count the entire diatom endosymbiont as the chloroplast, or just the red algal derived chloroplast inside it). The diatom endosymbiont has been reduced relatively little—it still retains its original mitochondria,and has endoplasmic reticulum, ribosomes, a nucleus, and of course, red algal derived chloroplasts—practically a complete cell, all inside the host's endoplasmic reticulum lumen. However the diatom endosymbiont can't store its own food—its starch is found in granules in the dinophyte host's cytoplasm instead. The diatom endosymbiont's nucleus is present, but it probably can't be called a nucleomorph because it shows no sign of genome reduction, and might have even been *expanded*. Diatoms have been engulfed by dinoflagellates at least three times.

The diatom endosymbiont is bounded by a single membrane, inside it are chloroplasts with four membranes. Like the diatom endosymbiont's diatom ancestor, the chloroplasts have triplet thylakoids and pyrenoids.

In some of these genera, the diatom endosymbiont's chloroplasts aren't the only chloroplasts in the dinophyte. The original three-membraned peridinin chloroplast is still around, converted to an eyespot.

Prasinophyte (Green Algal) Derived Dinophyte Chloroplast

Lepidodinium viride and its close relatives are dinophytes that lost their original peridinin chloroplast and replaced it with a green algal derived chloroplast (more specifically, a prasinophyte). *Lepidodinium* is the only dinophyte that has a chloroplast that's not from the rhodoplast lineage. The chloroplast is surrounded by two membranes and has no nucleomorph—all the nucleomorph genes have been transferred to the dinophyte nucleus. The endosymbiotic event that led to this chloroplast was serial secondary endosymbiosis rather than tertiary endosymbiosis—the endosymbiont was a green alga containing a primary chloroplast (making a secondary chloroplast).

Chromatophores

While most chloroplasts originate from that first set of endosymbiotic events, *Paulinella chromatophora* is an exception that acquired a photosynthetic cyanobacterial endosymbiont more recently. It is not clear whether that symbiont is closely related to the ancestral chloroplast of other eukaryotes. Being in the early stages of endosymbiosis, *Paulinella chromatophora* can offer some insights into how chloroplasts evolved. *Paulinella* cells contain one or two sausage shaped blue-green photosynthesizing structures called chromatophores, descended from the cyanobacterium *Synechococcus*. Chromatophores cannot survive outside their host. Chromatophore DNA is about a million base pairs long, containing around 850 protein encoding genes—far less than the three million base pair *Synechococcus* genome, but much larger than the approximately 150,000 base pair genome of the more assimilated chloroplast. Chromatophores have transferred much less of their DNA to the nucleus of their host. About 0.3–0.8% of the nuclear DNA in *Paulinella* is from the chromatophore, compared with 11–14% from the chloroplast in plants.

Kleptoplastidy

In some groups of mixotrophic protists, like some dinoflagellates, chloroplasts are separated from a captured alga or diatom and used temporarily. These klepto chloroplasts may only have a lifetime of a few days and are then replaced.

CHLOROPLAST DNA

Chloroplasts have their own DNA, often abbreviated as ctDNA, or cpDNA. It is also known as the plastome. Its existence was first proved in 1962, and first sequenced in 1986—when two Japanese research teams sequenced the chloroplast DNA of liverwort and tobacco. Since then, hundreds of chloroplast DNAs from various species have been sequenced, but they're mostly those of land plants and green algae—glaucophytes, red algae, and other algal groups are extremely underrepresented, potentially introducing some bias in views of "typical" chloroplast DNA structure and content.

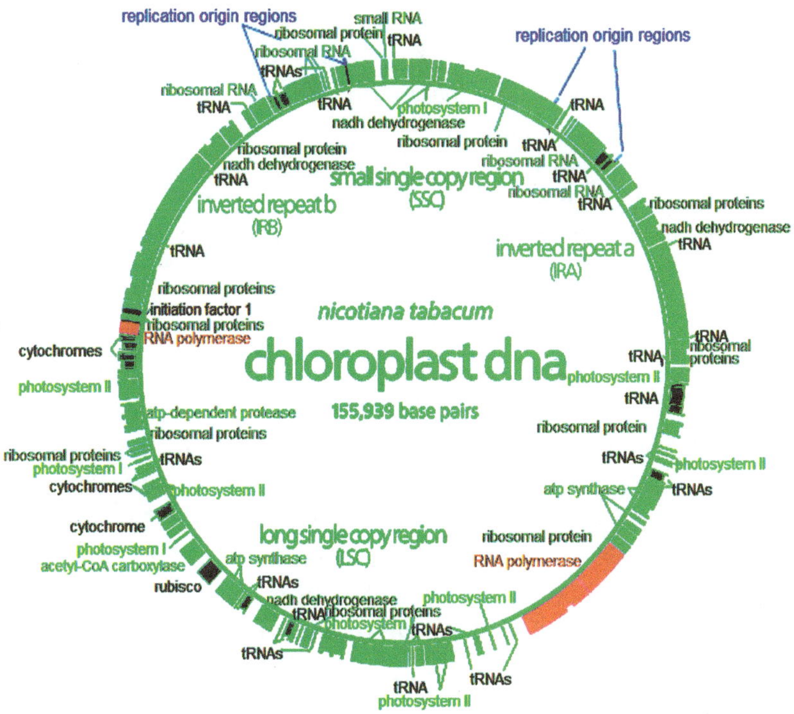

Chloroplast DNA Interactive gene map of chloroplast DNA from Nicotiana tabacum. Segments with labels on the inside reside on the B strand of DNA, segments with labels on the outside are on the A strand. Notches indicate introns.

Molecular Structure

With few exceptions, most chloroplasts have their entire chloroplast genome combined into a single large circular DNA molecule, typically 120,000–170,000 base pairs long. They can have a contour length of around 30–60 micrometers, and have a mass of about 80–130 million daltons.

While usually thought of as a circular molecule, there is some evidence that chloroplast DNA molecules more often take on a linear shape.

Inverted Repeats

Many chloroplast DNAs contain two *inverted repeats*, which separate a long single copy section (LSC) from a short single copy section (SSC). While a given pair of inverted repeats are rarely completely identical, they are always very similar to each other, apparently resulting from concerted evolution.

The inverted repeats vary wildly in length, ranging from 4,000 to 25,000 base pairs long each and containing as few as four or as many as over 150 genes.Inverted repeats in plants tend to be at the upper end of this range, each being 20,000–25,000 base pairs long.

The inverted repeat regions are highly conserved among land plants, and accumulate few mutations. Similar inverted repeats exist in the genomes of cyanobacteria and the other two chloroplast lineages (glaucophyta and rhodophyceae), suggesting that they predate the chloroplast, though some chloroplast DNAs have since lost or flipped the inverted repeats (making them direct repeats). It is possible that the inverted repeats help stabilize the rest of the chloroplast genome, as chloroplast DNAs which have lost some of the inverted repeat segments tend to get rearranged more.

Nucleoids

New chloroplasts may contain up to 100 copies of their DNA, though the number of chloroplast DNA copies decreases to about 15–20 as the chloroplasts age.They are usually packed into nucleoids, which can contain several identical chloroplast DNA rings. Many nucleoids can be found in each chloroplast. In primitive red algae, the chloroplast DNA nucleoids are clustered in the center of the chloroplast, while in green plants and green algae, the nucleoids are dispersed throughout the stroma.

Though chloroplast DNA is not associated with true histones, in red algae, similar proteins that tightly pack each chloroplast DNA ring into a nucleoid have been found.

DNA Replication

The Leading Model of cpDNA Replication

The mechanism for chloroplast DNA (cpDNA) replication has not been conclusively determined, but two main models have been proposed. Scientists have attempted to observe chloroplast replication via electron microscopy since the 1970s. The results of the microscopy experiments led to the idea that chloroplast DNA replicates using a double displacement loop (D-loop). As the D-loop moves through the circular DNA, it adopts a theta intermediary form, also known as a Cairns replication intermediate, and completes replication with a rolling circle mechanism. Transcription starts at specific points of origin. Multiple replication forks open up, allowing replication machinery to transcribe the DNA. As replication continues, the forks grow and eventually converge. The new cpDNA structures separate, creating daughter cpDNA chromosomes.

In addition to the early microscopy experiments, this model is also supported by the amounts of deamination seen in cpDNA. Deamination occurs when an amino group is lost and is a mutation that often results in base changes. When adenine is deaminated, it becomes hypoxanthine. Hypoxanthine can bind to cytosine, and when the XC base pair is replicated, it becomes a GC (thus, an A $\rightarrow$ G base change).

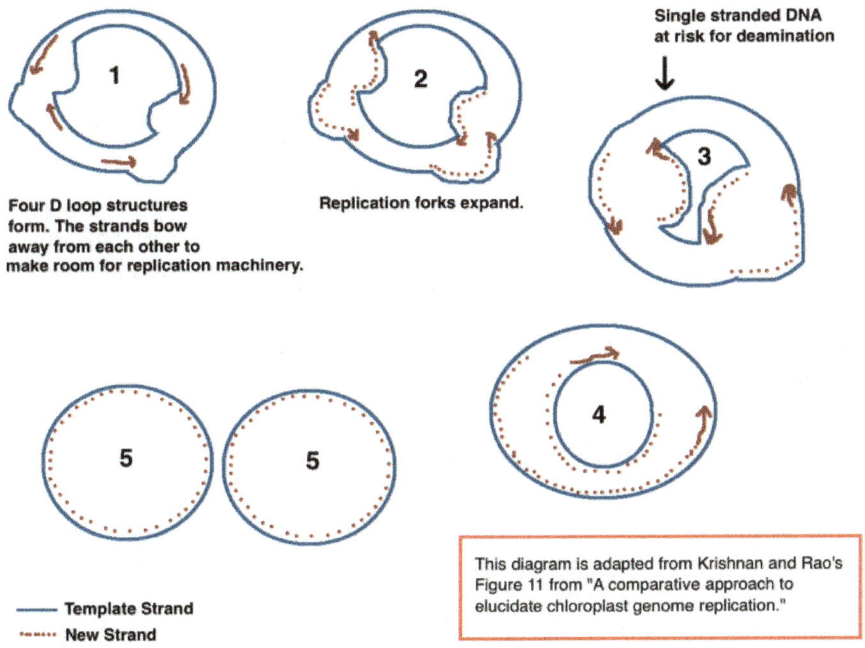

Chloroplast DNA replication via multiple D loop mechanisms. Adapted from Krishnan NM, Rao BJ's paper "A comparative approach to elucidate chloroplast genome replication."

Original DNA Strand

...CCATGCATGGATC...

Deamination of an Adenine

...CCATGCATGGATC...

↓

...CCHTGCATGGATC...

During Replication, H pairs with C

...CCHTGCATGGATC...
...GGCACGTACCTAG...

When Replicated Again, C pairs with G

...GGCACGTACCTAG...
...CCGTGCATGGATC...

Over time, base changes in the DNA sequence can arise from deamination mutations. When adenine is deaminated, it becomes hypoxanthine, which can pair with cytosine. During replication, the cytosine will pair with guanine, causing an A --> G base change.

Deamination

In cpDNA, there are several A → G deamination gradients. DNA becomes susceptible to deamination events when it is single stranded. When replication forks form, the strand not being copied is single stranded, and thus at risk for A → G deamination. Therefore, gradients in deamination indicate that replication forks were most likely present and the direction that they initially opened (the highest gradient is most likely nearest the start site because it was single stranded for the longest amount of time).This mechanism is still the leading theory today; however, a second theory suggests that most cpDNA is actually linear and replicates through homologous recombination. It further contends that only a minority of the genetic material is kept in circular chromosomes while the rest is in branched, linear, or other complex structures.

Alternative Model of Replication

One of competing model for cpDNA replication asserts that most cpDNA is linear and participates in homologous recombination and replication structures similar to bacteriophage T4. It has been established that some plants have linear cpDNA, such as maize, and that more species still contain complex structures that scientists do not yet understand. When the original experiments on cpDNA were performed, scientists did notice linear structures; however, they attributed these linear forms to broken circles. If the branched and complex structures seen in cpDNA experiments are real and not artifacts of concatenated circular DNA or broken circles, then a D-loop mechanism of replication is insufficient to explain how those structures would replicate. At the same time, homologous recombination does not expand the multiple A --> G gradients seen in plastomes. Because of the failure to explain the deamination gradient as well as the numerous plant species that have been shown to have circular cpDNA, the predominant theory continues to hold that most cpDNA is circular and most likely replicates via a D loop mechanism.

Gene Content and Protein Synthesis

The chloroplast genome most commonly includes around 100 genes that code for a variety of things, mostly to do with the protein pipeline and photosynthesis. As in prokaryotes, genes in chloroplast DNA are organized into operons. Interestingly though, unlike prokaryotic DNA molecules, chloroplast DNA molecules contain introns (plant mitochondrial DNAs do too, but not human mtDNAs).

Among land plants, the contents of the chloroplast genome are fairly similar.

Chloroplast Genome Reduction and Gene Transfer

Over time, many parts of the chloroplast genome were transferred to the nuclear genome of the host, a process called *endosymbiotic gene transfer*. As a result, the chloroplast genome is heavily reduced compared to that of free-living cyanobacteria. Chloroplasts may contain 60–100 genes whereas cyanobacteria often have more than 1500 genes in their genome. Recently, a plastid without a genome was found, demonstrating chloroplasts can lose their genome during endosymbiotic the gene transfer process.

Endosymbiotic gene transfer is how we know about the lost chloroplasts in many chromalveolate lineages. Even if a chloroplast is eventually lost, the genes it donated to the former host's nucleus persist, providing evidence for the lost chloroplast's existence. For example, while diatoms (a heterokontophyte) now have a red algal derived chloroplast, the presence of many green algal genes in the diatom nucleus provide evidence that the diatom ancestor (probably the ancestor of all chromalveolates too) had a green algal derived chloroplast at some point, which was subsequently replaced by the red chloroplast.

In land plants, some 11–14% of the DNA in their nuclei can be traced back to the chloroplast, up to 18% in *Arabidopsis*, corresponding to about 4,500 protein-coding genes. There have been a few recent transfers of genes from the chloroplast DNA to the nuclear genome in land plants.

Of the approximately 3000 proteins found in chloroplasts, some 95% of them are encoded by nuclear genes. Many of the chloroplast's protein complexes consist of subunits from both the chloroplast genome and the host's nuclear genome. As a result, protein synthesis must be coordinated between the chloroplast and the nucleus. The chloroplast is mostly under nuclear control, though chloroplasts can also give out signals regulating gene expression in the nucleus, called *retrograde signaling*.

Protein Synthesis

Protein synthesis within chloroplasts relies on two RNA polymerases. One is coded by the chloroplast DNA, the other is of nuclear origin. The two RNA polymerases may recognize and bind to different kinds of promoters within the chloroplast genome. The ribosomes in chloroplasts are similar to bacterial ribosomes.

Protein Targeting and Import

Because so many chloroplast genes have been moved to the nucleus, many proteins that would originally have been translated in the chloroplast are now synthesized in the cytoplasm of the plant cell. These proteins must be directed back to the chloroplast, and imported through at least two chloroplast membranes.

Curiously, around half of the protein products of transferred genes aren't even targeted back to the chloroplast. Many became exaptations, taking on new functions like participating in cell division, protein routing, and even disease resistance. A few chloroplast genes found new homes in the mitochondrial genome—most became nonfunctional pseudogenes, though a few tRNA genes still work in the mitochondrion. Some transferred chloroplast DNA protein products get directed to the secretory pathway though it should be noted that many secondary plastids are bounded by an outermost membrane derived from the host's cell membrane, and therefore topologically outside of the cell, because to reach the chloroplast from the cytosol, you have to cross the cell membrane, just like if you were headed for the extracellular space. In those cases, chloroplast-targeted proteins do initially travel along the secretory pathway.

Because the cell acquiring a chloroplast already had mitochondria (and peroxisomes, and a cell membrane for secretion), the new chloroplast host had to develop a unique protein targeting system to avoid having chloroplast proteins being sent to the wrong organelle.

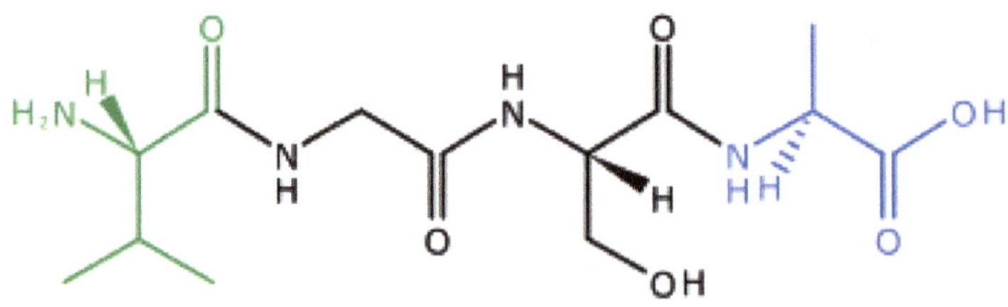

The two ends of a polypeptide are called the N-terminus, or *amino end*, and the C-terminus, or *carboxyl end*. This polypeptide has four amino acids linked together. At the left is the N-terminus, with its amino (H_2N) group in green. The blue C-terminus, with its carboxyl group (CO_2H) is at the right.

In most, but not all cases, nuclear-encoded chloroplast proteins are translated with a *cleavable transit peptide* that's added to the N-terminus of the protein precursor. Sometimes the transit sequence is found on the C-terminus of the protein, or within the functional part of the protein.

Transport Proteins and Membrane Translocons

After a chloroplast polypeptide is synthesized on a ribosome in the cytosol, an enzyme specific to chloroplast proteins phosphorylates, or adds a phosphate group to many (but not all) of them in their transit sequences. Phosphorylation helps many proteins bind the polypeptide, keeping it from folding prematurely. This is important because it prevents chloroplast proteins from assuming their active form and carrying out their chloroplast functions in the wrong place—the cytosol. At the same time, they have to keep just enough shape so that they can be recognized by the chloroplast. These proteins also help the polypeptide get imported into the chloroplast.

From here, chloroplast proteins bound for the stroma must pass through two protein complexes—the TOC complex, or *translocon on the* ***o****uter chloroplast membrane*, and the TIC translocon, or *translocon on the* ***i****nner chloroplast membrane translocon*. Chloroplast polypeptide chains probably often travel through the two complexes at the same time, but the TIC complex can also retrieve preproteins lost in the intermembrane space.

Structure

In land plants, chloroplasts are generally lens-shaped, 5–8 μm in diameter and 1–3 μm thick. Greater diversity in chloroplast shapes exists among the algae, which often contain a single chloroplast that can be shaped like a net (e.g., *Oedogonium*), a cup (e.g., *Chlamydomonas*), a ribbon-like spiral around the edges of the cell (e.g., *Spirogyra*), or slightly twisted bands at the cell edges (e.g., *Sirogonium*). Some algae have two chloroplasts in each cell; they are star-shaped in *Zygnema*, or may follow the shape of half the cell in order Desmidiales. In some algae, the chloroplast takes up most of the cell, with pockets for the nucleus and other organelles, for example, some species of *Chlorella* have a cup-shaped chloroplast that occupies much of the cell.

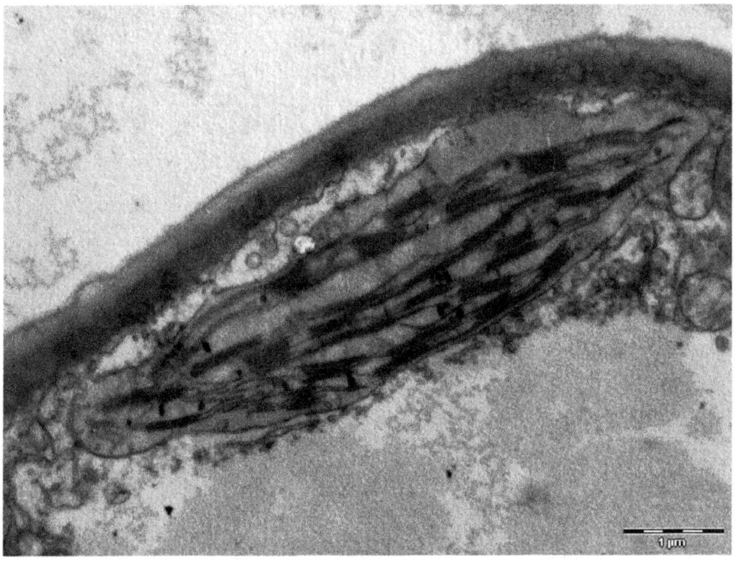

Transmission electron microscope image of a chloroplast. Grana of thylakoids and their connecting lamellae are clearly visible.

All chloroplasts have at least three membrane systems—the outer chloroplast membrane, the inner chloroplast membrane, and the thylakoid system. Chloroplasts that are the product of secondary endosymbiosis may have additional membranes surrounding these three. Inside the outer and inner chloroplast membranes is the chloroplast stroma, a semi-gel-like fluid that makes up much of a chloroplast›s volume, and in which the thylakoid system floats.

There are some common misconceptions about the outer and inner chloroplast membranes. The fact that chloroplasts are surrounded by a double membrane is often cited as evidence that they are the descendants of endosymbiotic cyanobacteria. This is often interpreted as meaning the outer chloroplast membrane is the product of the host's cell membrane infolding to form a vesicle to surround the ancestral cyanobacterium—which is not true—both chloroplast membranes are homologous to the cyanobacterium's original double membranes.

The chloroplast double membrane is also often compared to the mitochondrial double membrane. This is not a valid comparison—the inner mitochondria membrane is used to run proton pumps and carry out oxidative phosphorylation across to generate ATP energy. The only chloroplast structure that can considered analogous to it is the internal thylakoid system. Even so, in terms of "in-out", the direction of chloroplast H+ ion flow is in the opposite direction compared to oxidative phosphorylation in mitochondria. In addition, in terms of function, the inner chloroplast membrane, which regulates metabolite passage and synthesizes some materials, has no counterpart in the mitochondrion.

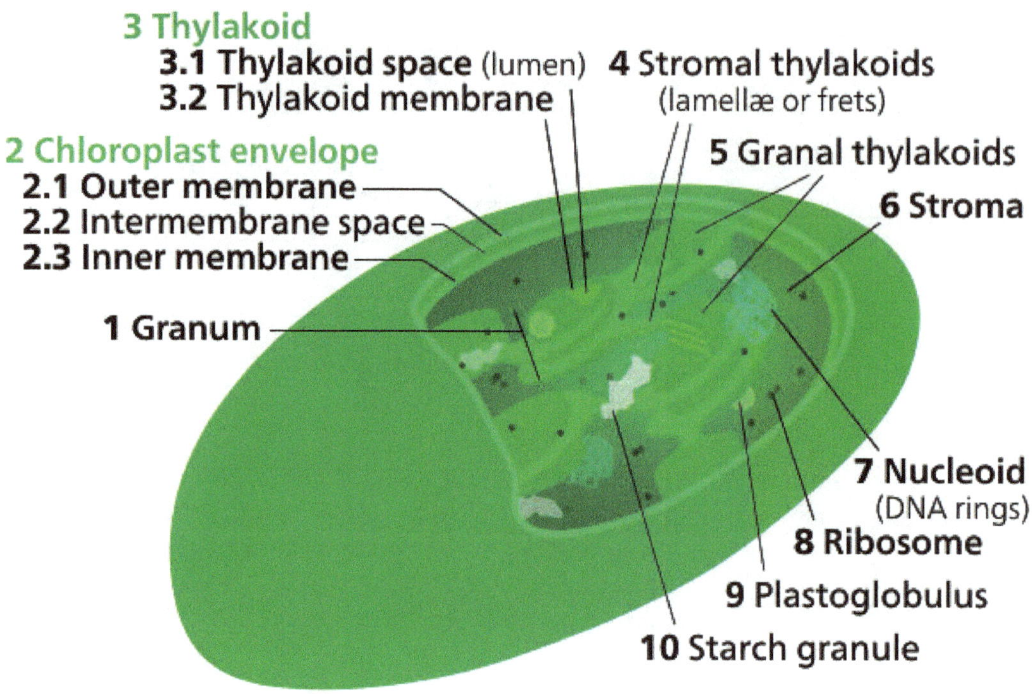

Chloroplast ultrastructure (interactive diagram) Chloroplasts have at least three distinct membrane systems, and a variety of things can be found in their stroma.

Outer Chloroplast Membrane

The outer chloroplast membrane is a semi-porous membrane that small molecules and ions can easily diffuse across. However, it is not permeable to larger proteins, so chloroplast polypeptides being synthesized in the cell cytoplasm must be transported across the outer chloroplast membrane by the TOC complex, or ***t**ranslocon on the **o**uter **c**hloroplast* membrane.

The chloroplast membranes sometimes protrude out into the cytoplasm, forming a stromule, or **stroma**-containing tub**ule**. Stromules are very rare in chloroplasts, and are much more common in other plastids like chromoplasts and amyloplasts in petals and roots, respectively. They may exist to increase the chloroplast's surface area for cross-membrane transport, because they are often branched and tangled with the endoplasmic reticulum. When they were first observed in 1962, some plant biologists dismissed the structures as artifactual, claiming that stromules were just oddly shaped chloroplasts with constricted regions or dividing chloroplasts. However, there is a growing body of evidence that stromules are functional, integral features of plant cell plastids, not merely artifacts.

Intermembrane Space and Peptidoglycan Wall

Instead of an intermembrane space, glaucophyte algae have a peptidoglycan wall between their inner and outer chloroplast membranes.

Usually, a thin intermembrane space about 10–20 nanometers thick exists between the outer and inner chloroplast membranes.

Glaucophyte algal chloroplasts have a peptidoglycan layer between the chloroplast membranes. It corresponds to the peptidoglycan cell wall of their cyanobacterial ancestors, which is located between their two cell membranes. These chloroplasts are called *muroplasts* (from Latin *«mura»*, meaning "wall"). Other chloroplasts have lost the cyanobacterial wall, leaving an intermembrane space between the two chloroplast envelope membranes.

Inner Chloroplast Membrane

The inner chloroplast membrane borders the stroma and regulates passage of materials in and out of the chloroplast. After passing through the TOC complex in the outer chloroplast membrane, polypeptides must pass through the TIC complex*(translocon on the **i**nner chloroplast membrane)* which is located in the inner chloroplast membrane.

In addition to regulating the passage of materials, the inner chloroplast membrane is where fatty acids, lipids, and carotenoids are synthesized.

Peripheral Reticulum

Some chloroplasts contain a structure called the chloroplast peripheral reticulum. It is often found in the chloroplasts of C_4 plants, though it has also been found in some C_3

angiosperms, and even some gymnosperms. The chloroplast peripheral reticulum consists of a maze of membranous tubes and vesicles continuous with the inner chloroplast membrane that extends into the internal stromal fluid of the chloroplast. Its purpose is thought to be to increase the chloroplast's surface area for cross-membrane transport between its stroma and the cell cytoplasm. The small vesicles sometimes observed may serve as transport vesicles to shuttle stuff between the thylakoids and intermembrane space.

Stroma

The protein-rich, alkaline, aqueous fluid within the inner chloroplast membrane and outside of the thylakoid space is called the stroma, which corresponds to the cytosol of the original cyanobacterium. Nucleoids of chloroplast DNA, chloroplast ribosomes, the thylakoid system with plastoglobuli, starch granules, and many proteins can be found floating around in it. The Calvin cycle, which fixes CO_2 into sugar takes place in the stroma.

Chloroplast Ribosomes

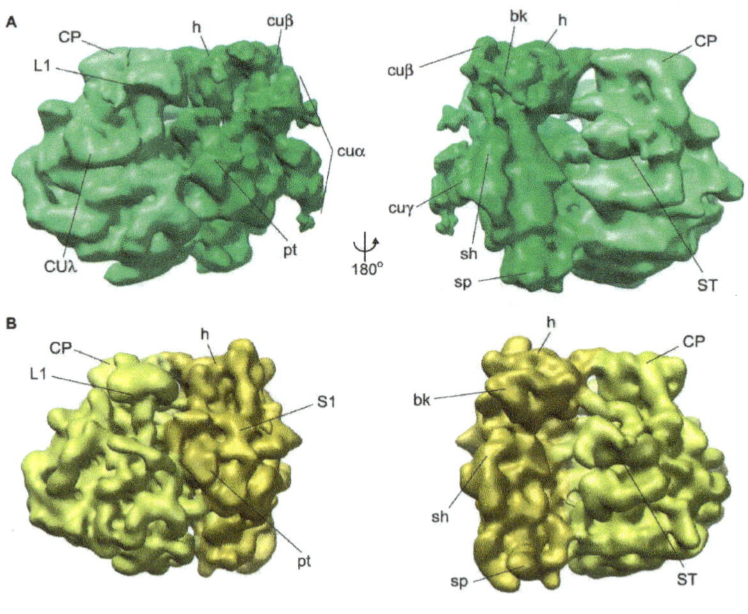

Chloroplast ribosomes Comparison of a chloroplast ribosome (green) and a bacterial ribosome (yellow). Important features common to both ribosomes and chloroplast-unique features are labeled.

Chloroplasts have their own ribosomes, which they use to synthesize a small fraction of their proteins. Chloroplast ribosomes are about two-thirds the size of cytoplasmic ribosomes (around 17 nm vs 25 nm). They take mRNAs transcribed from the chloroplast DNA and translate them into protein. While similar to bacterial ribosomes,chloroplast translation is more complex than in bacteria, so chloroplast ribosomes include some chloroplast-unique features. Small subunit ribosomal RNAs in several Chlorophyta and euglenid chloroplasts lack motifs for shine-dalgarno sequence recognition, which is considered essential for translation initiation in most chloroplasts and prokaryotes. Such loss is also rarely observed in other plastids and prokaryotes.

Plastoglobuli

Plastoglobuli (**singular** *plastoglobulus*, sometimes spelled *plastoglobule(s)*), are spherical bubbles of lipids and proteins about 45–60 nanometers across. They are surrounded by a lipid monolayer. Plastoglobuli are found in all chloroplasts, but become more common when the chloroplast is under oxidative stress, or when it ages and transitions into a gerontoplast. Plastoglobuli also exhibit a greater size variation under these conditions. They are also common in etioplasts, but decrease in number as the etioplasts mature into chloroplasts.

Plastoglubuli contain both structural proteins and enzymes involved in lipid synthesis and metabolism. They contain many types of lipids including plastoquinone, vitamin E, carotenoids and chlorophylls.

Plastoglobuli were once thought to be free-floating in the stroma, but it is now thought that they are permanently attached either to a thylakoid or to another plastoglobulus attached to a thylakoid, a configuration that allows a plastoglobulus to exchange its contents with the thylakoid network. In normal green chloroplasts, the vast majority of plastoglobuli occur singularly, attached directly to their parent thylakoid. In old or stressed chloroplasts, plastoglobuli tend to occur in linked groups or chains, still always anchored to a thylakoid.

Plastoglobuli form when a bubble appears between the layers of the lipid bilayer of the thylakoid membrane, or bud from existing plastoglubuli—though they never detach and float off into the stroma. Practically all plastoglobuli form on or near the highly curved edges of the thylakoid disks or sheets. They are also more common on stromal thylakoids than on granal ones.

Starch Granules

Starch granules are very common in chloroplasts, typically taking up 15% of the organelle's volume, though in some other plastids like amyloplasts, they can be big enough to distort the shape of the organelle. Starch granules are simply accumulations of starch in the stroma, and are not bounded by a membrane.

Starch granules appear and grow throughout the day, as the chloroplast synthesizes sugars, and are consumed at night to fuel respiration and continue sugar export into the phloem, though in mature chloroplasts, it is rare for a starch granule to be completely consumed or for a new granule to accumulate.

Starch granules vary in composition and location across different chloroplast lineages. In red algae, starch granules are found in the cytoplasm rather than in the chloroplast. In C_4 plants, mesophyll chloroplasts, which do not synthesize sugars, lack starch granules.

Rubisco

Rubisco, shown here in a space-filling model, is the main enzyme responsible for carbon fixation in chloroplasts.

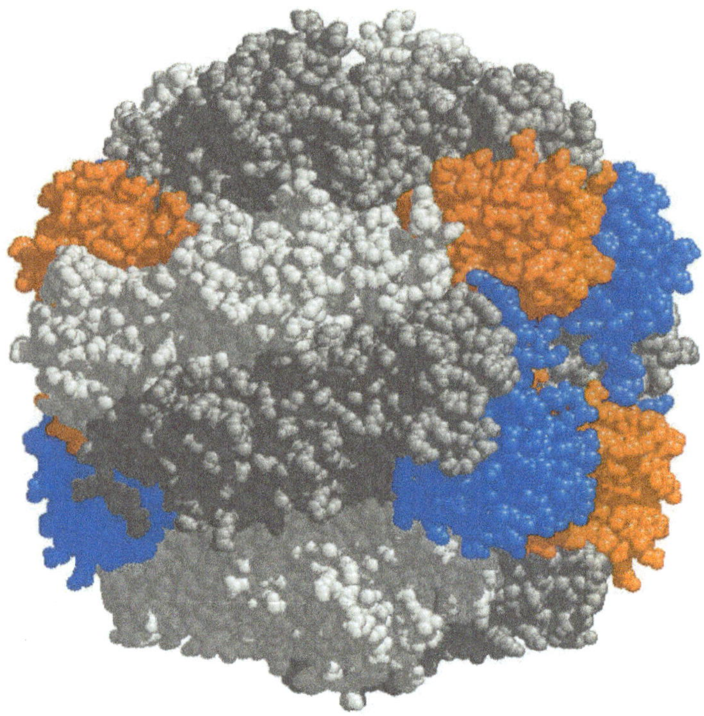

The chloroplast stroma contains many proteins, though the most common and important is Rubisco, which is probably also the most abundant protein on the planet. Rubisco is the enzyme that fixes CO_2 into sugar molecules. In C_3 plants, rubisco is abundant in all chloroplasts, though in C_4 plants, it is confined to the bundle sheath chloroplasts, where the Calvin cycle is carried out in C_4 plants.

Pyrenoids

The chloroplasts of some hornworts and algae contain structures called pyrenoids. They are not found in higher plants. Pyrenoids are roughly spherical and highly refractive bodies which are a site of starch accumulation in plants that contain them. They consist of a matrix opaque to electrons, surrounded by two hemispherical starch plates. The starch is accumulated as the pyrenoids mature. In algae with carbon concentrating mechanisms, the enzyme rubisco is found in the pyrenoids. Starch can also accumulate around the pyrenoids when CO_2 is scarce. Pyrenoids can divide to form new pyrenoids, or be produced "de novo".

Thylakoid System

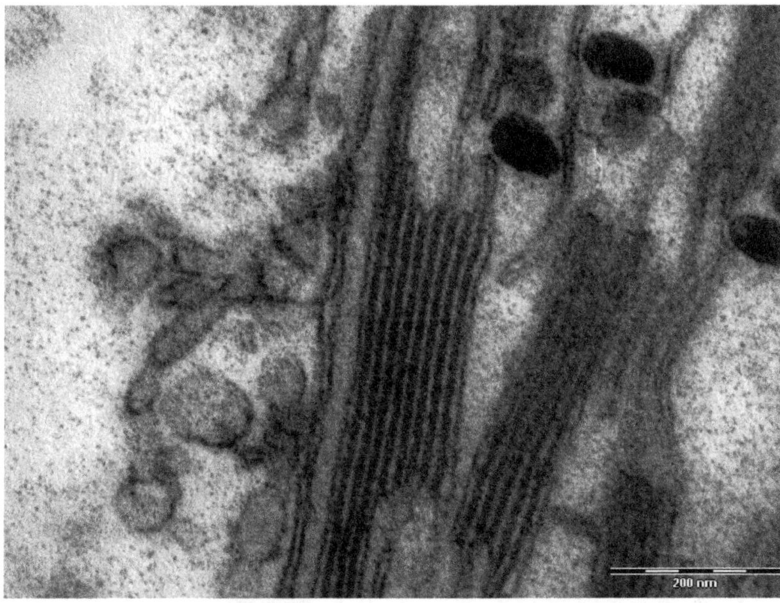

Transmission electron microscope image of some thylakoids arranged in grana stacks and lamellæ. Plastoglobuli (dark blobs) are also present.

Suspended within the chloroplast stroma is the thylakoid system, a highly dynamic collection of membranous sacks called thylakoids where chlorophyll is found and the light reactions of photosynthesis happen. In most vascular plant chloroplasts, the thylakoids are arranged in stacks called grana, though in certain C_4 plant chloroplasts and some algal chloroplasts, the thylakoids are free floating.

Granal Structure

Using a light microscope, it is just barely possible to see tiny green granules—which were named grana. With electron microscopy, it became possible to see the thylakoid system in more detail, revealing it to consist of stacks of flat thylakoids which made up the grana, and long interconnecting stromal thylakoids which linked different grana. In the transmission electron microscope, thylakoid membranes appear as alternating light-and-dark bands, 8.5 nanometers thick.

For a long time, the three-dimensional structure of the thylakoid system has been unknown or disputed. One model has the granum as a stack of thylakoids linked by helical stromal thylakoids; the other has the granum as a single folded thylakoid connected in a "hub and spoke" way to other grana by stromal thylakoids. While the thylakoid system is still commonly depicted according to the folded thylakoid model, it was determined in 2011 that the stacked and helical thylakoids model is correct.

Granum structure The prevailing model for granal structure is a stack of granal thylakoids linked by helical stromal thylakoids that wrap around the grana stacks and form large sheets that connect different grana.

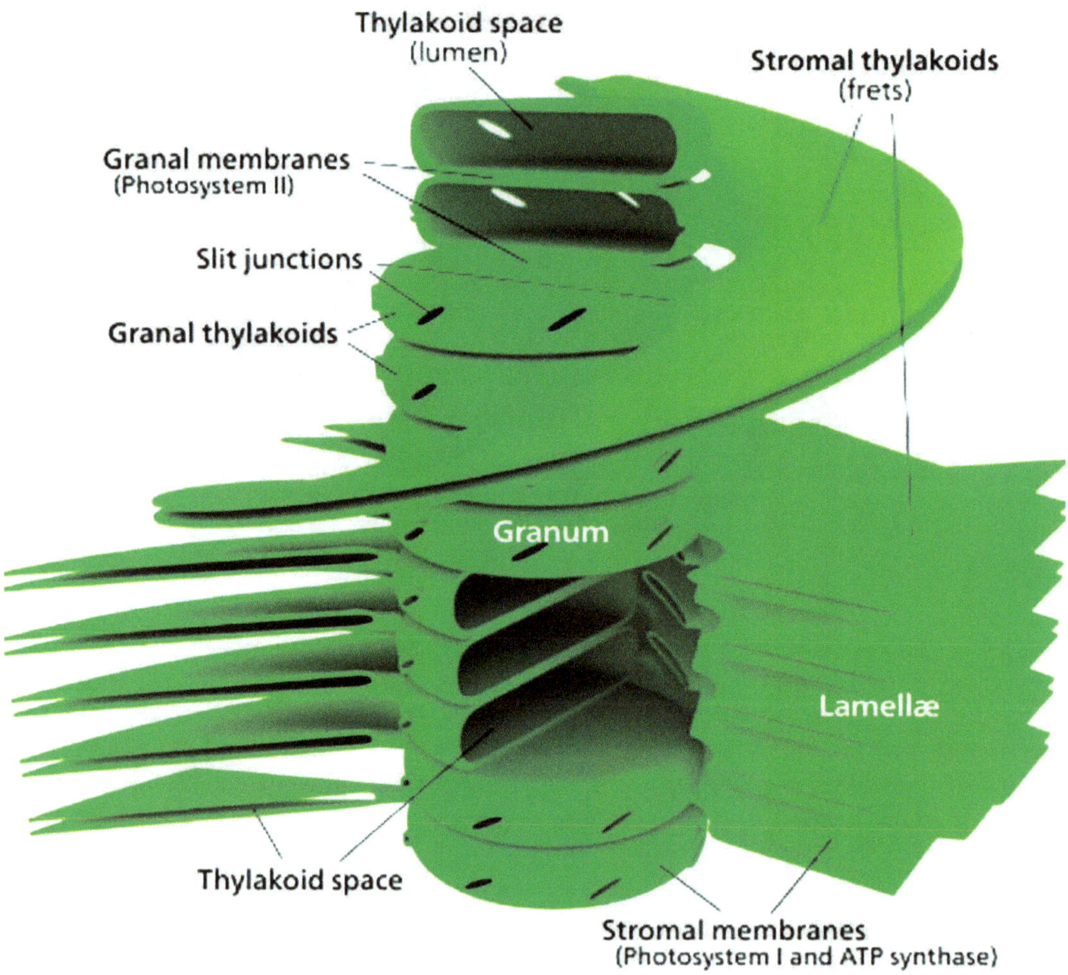

In the helical thylakoid model, grana consist of a stack of flattened circular granal thylakoids that resemble pancakes. Each granum can contain anywhere from two to a hundred thylakoids, though grana with 10–20 thylakoids are most common. Wrapped around the grana are helicoid stromal thylakoids, also known as frets or lamellar thylakoids. The helices ascend at an angle of 20–25°, connecting to each granal thylakoid at a bridge-like slit junction. The helicoids may extend as large sheets that link multiple grana, or narrow to tube-like bridges between grana. While different parts of the thylakoid system contain different membrane proteins, the thylakoid membranes are continuous and the thylakoid space they enclose form a single continuous labyrinth.

Thylakoids

Thylakoids (sometimes spelled *thylakoïds*), are small interconnected sacks which contain the membranes that the light reactions of photosynthesis take place on. The word *thylakoid* comes from the Greek word *thylakos* which means "sack".

Embedded in the thylakoid membranes are important protein complexes which carry out the light reactions of photosynthesis. Photosystem II and photosystem I contain light-harvesting complexes with chlorophyll and carotenoids that absorb light energy and use it to energize electrons. Molecules in the thylakoid membrane use the energized electrons to pump hydrogen ions into the thylakoid space, decreasing the pH and turning it acidic. ATP synthase is a large protein complex that harnesses the concentration gradient of the hydrogen ions in the thylakoid space to generate ATP energy as the hydrogen ions flow back out into the stroma—much like a dam turbine.

There are two types of thylakoids—granal thylakoids, which are arranged in grana, and stromal thylakoids, which are in contact with the stroma. Granal thylakoids are pancake-shaped circular disks about 300–600 nanometers in diameter. Stromal thylakoids are helicoid sheets that spiral around grana. The flat tops and bottoms of granal thylakoids contain only the relatively flat photosystem II protein complex. This allows them to stack tightly, forming grana with many layers of tightly appressed membrane, called granal membrane, increasing stability and surface area for light capture.

In contrast, photosystem I and ATP synthase are large protein complexes which jut out into the stroma. They can›t fit in the appressed granal membranes, and so are found in the stromal thylakoid membrane—the edges of the granal thylakoid disks and the stromal thylakoids. These large protein complexes may act as spacers between the sheets of stromal thylakoids.

The number of thylakoids and the total thylakoid area of a chloroplast is influenced by light exposure. Shaded chloroplasts contain larger and more grana with more thylakoid membrane area than chloroplasts exposed to bright light, which have smaller and fewer grana and less thylakoid area. Thylakoid extent can change within minutes of light exposure or removal.

Pigments and Chloroplast Colors

Inside the photosystems embedded in chloroplast thylakoid membranes are various photosynthetic pigments, which absorb and transfer light energy. The types of pigments found are different in various groups of chloroplasts, and are responsible for a wide variety of chloroplast colorations.

Chlorophylls

Chlorophyll *a* is found in all chloroplasts, as well as their cyanobacterial ancestors. Chlorophyll *a* is a blue-green pigment partially responsible for giving most cyanobacteria and chloroplasts their color. Other forms of chlorophyll exist, such as the accessory pigments chlorophyll *b*, chlorophyll *c*, chlorophyll *d*, and chlorophyll *f*.

Chlorophyll *b* is an olive green pigment found only in the chloroplasts of plants, green algae, any secondary chloroplasts obtained through the secondary endosymbiosis of a green alga, and a few cyanobacteria. It is the chlorophylls *a* and *b* together that make most plant and green algal chloroplasts green.

Chlorophyll *c* is mainly found in secondary endosymbiotic chloroplasts that originated from a red alga, although it is not found in chloroplasts of red algae themselves. Chlorophyll *c* is also found in some green algae and cyanobacteria.

Chlorophylls *d* and *f* are pigments found only in some cyanobacteria.

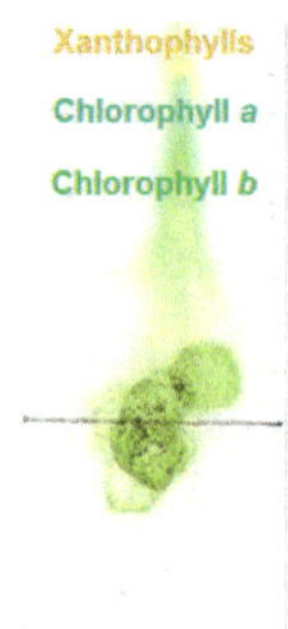

Carotenoids

Delesseria sanguinea, a red alga, has chloroplasts that contain red pigments like phycoerytherin that mask their blue-green chlorophyll *a*.

In addition to chlorophylls, another group of yellow–orange pigments called carotenoids are also found in the photosystems. There are about thirty photosynthetic carotenoids. They help transfer and dissipate excess energy, and their bright colors sometimes override the chlorophyll green, like during the fall, when the leaves of some land plants change color. β-carotene is a bright red-orange carotenoid found in nearly all chloroplasts, like chlorophyll *a*.Xanthophylls, especially the orange-red zeaxanthin, are also common. Many other forms of carotenoids exist that are only found in certain groups of chloroplasts.

Phycobilins

Phycobilins are a third group of pigments found in cyanobacteria, and glaucophyte, red algal, and cryptophyte chloroplasts. Phycobilins come in all colors, though phycoerytherin is one of the pigments that makes many red algae red. Phycobilins often organize into relatively large protein complexes about 40 nanometers across called phycobilisomes. Like photosystem I and ATP synthase, phycobilisomes jut into the stroma, preventing thylakoid stacking in red algal chloroplasts.Cryptophyte chloroplasts and some cyanobacteria don't have their phycobilin pigments organized into phycobilisomes, and keep them in their thylakoid space instead.

Photosynthetic pigments Table of the presence of various pigments across chloroplast groups. Colored cells represent pigment presence.

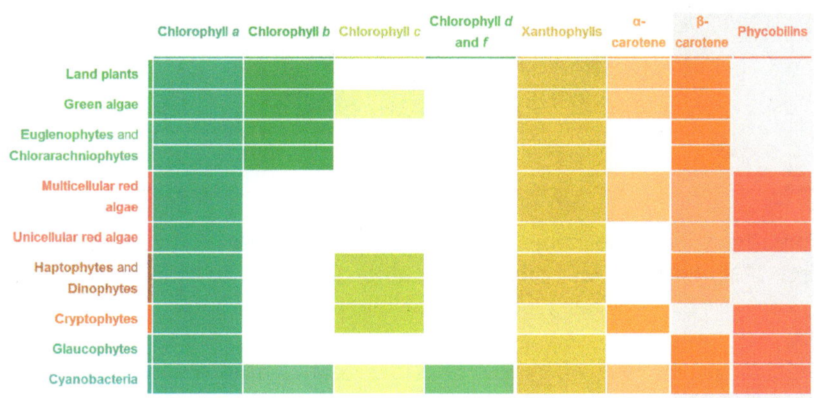

Specialized Chloroplasts in C_4 Plants

To fix carbon dioxide into sugar molecules in the process of photosynthesis, chloroplasts use an enzyme called rubisco. Rubisco has a problem—it has trouble distinguishing between carbon dioxide and oxygen, so at high oxygen concentrations, rubisco starts accidentally adding oxygen to sugar precursors. This has the end result of ATP energy being wasted and CO_2 being released, all with no sugar being produced. This is a big problem, since O_2 is produced by the initial light reactions of photosynthesis, causing issues down the line in the Calvin cycle which uses rubisco.

C_4 plants evolved a way to solve this—by spatially separating the light reactions and the Calvin cycle. The light reactions, which store light energy in ATP and NADPH, are done in the mesophyll cells of a C_4 leaf. The Calvin cycle, which uses the stored energy to make sugar using rubisco, is done in the bundle sheath cells, a layer of cells surrounding a vein in a leaf.

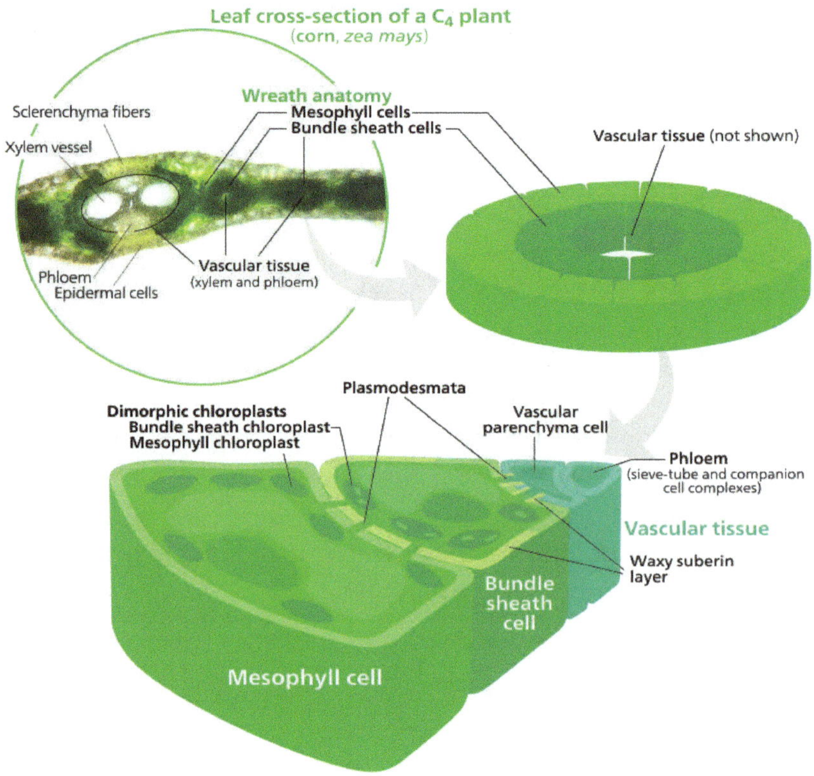

Many C_4 plants have their mesophyll cells and bundle sheath cells arranged radially around their leaf veins. The two types of cells contain different types of chloroplasts specialized for a particular part of photosynthesis.

As a result, chloroplasts in C_4 mesophyll cells and bundle sheath cells are specialized for each stage of photosynthesis. In mesophyll cells, chloroplasts are specialized for the light reactions, so they lack rubisco, and have normal grana and thylakoids, which they use to make ATP and NADPH, as well as oxygen. They store CO_2 in a four-carbon compound, which is why the process is called *C_4 photosynthesis*. The four-carbon compound is then transported to the bundle sheath chloroplasts, where it drops off CO_2 and returns to the mesophyll. Bundle sheath chloroplasts do not carry out the light reactions, preventing oxygen from building up in them and disrupting rubisco activity. Because of this, they lack thylakoids organized into grana stacks—though bundle sheath chloroplasts still have free-floating thylakoids in the stroma where they still carry out cyclic electron flow, a light-driven method of synthesizing ATP to power the Calvin cycle without generating oxygen. They lack photosystem II, and only have photosystem I—the only protein complex needed for cyclic electron flow. Because the job of bundle sheath chloroplasts is to carry out the Calvin cycle and make sugar, they often contain large starch grains.

Both types of chloroplast contain large amounts of chloroplast peripheral reticulum, which they use to get more surface area to transport stuff in and out of them. Mesophyll chloroplasts have a little more peripheral reticulum than bundle sheath chloroplasts.

Location

Distribution in a Plant

Not all cells in a multicellular plant contain chloroplasts. All green parts of a plant contain chloroplasts—the chloroplasts, or more specifically, the chlorophyll in them are what make the photosynthetic parts of a plant green. The plant cells which contain chloroplasts are usually parenchyma cells, though chloroplasts can also be found in collenchyma tissue. A plant cell which contains chloroplasts is known as a chlorenchyma cell. A typical chlorenchyma cell of a land plant contains about 10 to 100 chloroplasts.

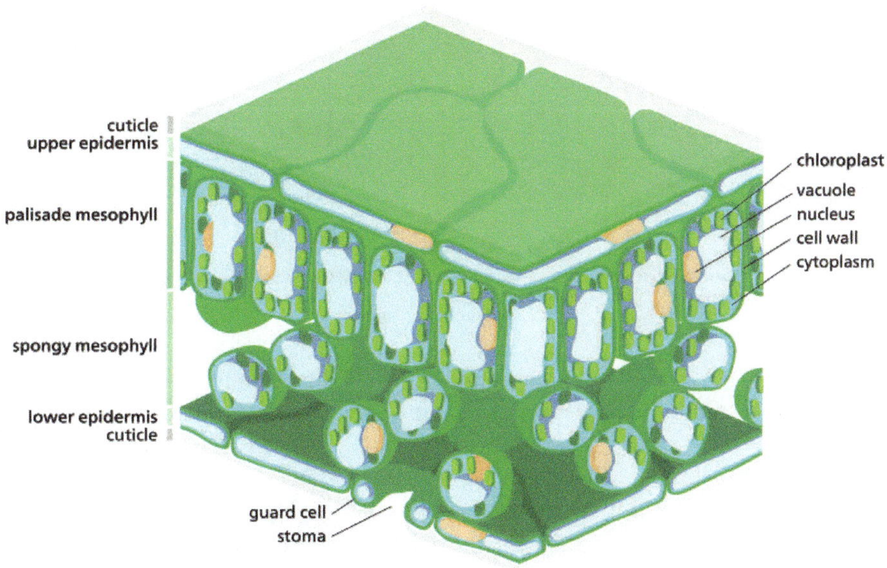

A cross section of a leaf, showing chloroplasts in its mesophyll cells. Stomal guard cells also have chloroplasts, though much fewer than mesophyll cells.

In some plants such as cacti, chloroplasts are found in the stems, though in most plants, chloroplasts are concentrated in the leaves. One square millimeter of leaf tissue can contain half a million chloroplasts. Within a leaf, chloroplasts are mainly found in the mesophyll layers of a leaf, and the guard cells of stomata. Palisade mesophyll cells can contain 30–70 chloroplasts per cell, while stomatal guard cells contain only around 8–15 per cell, as well as much less chlorophyll. Chloroplasts can also be found in the bundle sheath cells of a leaf, especially in C_4 plants, which carry out the Calvin cycle in their bundle sheath cells. They are often absent from the epidermis of a leaf.

Cellular Location

Chloroplast Movement

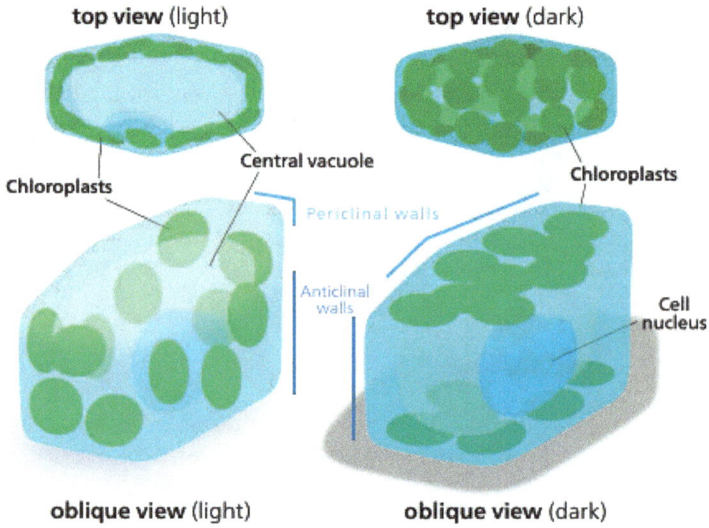

When chloroplasts are exposed to direct sunlight, they stack along the anticlinal cell walls to minimize exposure. In the dark they spread out in sheets along the periclinal walls to maximize light absorption.

The chloroplasts of plant and algal cells can orient themselves to best suit the available light. In low-light conditions, they will spread out in a heet—maximizing the surface area to absorb light. Under intense light, they will seek shelter by aligning in vertical columns along the plant cell's cell wall or turning sideways so that light strikes them edge-on. This reduces exposure and protects them from photooxidative damage. This ability to distribute chloroplasts so that they can take shelter behind each other or spread out may be the reason why land plants evolved to have many small chloroplasts instead of a few big ones. Chloroplast movement is considered one of the most closely regulated stimulus-response systems that can be found in plants. Mitochondria have also been observed to follow chloroplasts as they move.

In higher plants, chloroplast movement is run by phototropins, blue light photoreceptors also responsible for plant phototropism. In some algae, mosses, ferns, and flowering plants, chloroplast movement is influenced by red light in addition to blue light, though very long red wavelengths inhibit movement rather than speeding it up. Blue light generally causes chloroplasts to seek shelter, while red light draws them out to maximize light absorption.

Studies of *Vallisneria gigantea*, an aquatic flowering plant, have shown that chloroplasts can get moving within five minutes of light exposure, though they don't initially show any net directionality. They may move along microfilament tracks, and the fact that the microfilament mesh changes shape to form a honeycomb structure surrounding the chloroplasts after they have moved suggests that microfilaments may help to anchor chloroplasts in place.

FUNCTION AND CHEMISTRY

Guard Cell Chloroplasts

Unlike most epidermal cells, the guard cells of plant stomata contain relatively well-developed chloroplasts. However, exactly what they do is controversial.

Plant Innate Immunity

Plants lack specialized immune cells—all plant cells participate in the plant immune response. Chloroplasts, along with the nucleus, cell membrane, and endoplasmic reticulum, are key players in pathogen defense. Due to its role in a plant cell's immune response, pathogens frequently target the chloroplast.

Plants have two main immune responses—the hypersensitive response, in which infected cells seal themselves off and undergo programmed cell death, and systemic acquired resistance, where infected cells release signals warning the rest of the plant of a pathogen's presence. Chloroplasts stimulate both responses by purposely damaging their photosynthetic system, producing reactive oxygen species. High levels of reactive oxygen species will cause the hypersensitive response. The reactive oxygen species also directly kill any pathogens within the cell. Lower levels of reactive oxygen species initiate systemic acquired resistance, triggering defense-molecule production in the rest of the plant.

In some plants, chloroplasts are known to move closer to the infection site and the nucleus during an infection.

Chloroplasts can serve as cellular sensors. After detecting stress in a cell, which might be due to a pathogen, chloroplasts begin producing molecules like salicylic acid, jasmonic acid, nitric oxide and reactive oxygen species which can serve as defense-signals. As cellular signals, reactive oxygen species are unstable molecules, so they probably don't leave the chloroplast, but instead pass on their signal to an unknown second messenger molecule. All these molecules initiate retrograde signaling—signals from the chloroplast that regulate gene expression in the nucleus.

In addition to defense signaling, chloroplasts, with the help of the peroxisomes, help synthesize an important defense molecule, jasmonate. Chloroplasts synthesize all the fatty acids in a plant cell—linoleic acid, a fatty acid, is a precursor to jasmonate.

Photosynthesis

One of the main functions of the chloroplast is its role in photosynthesis, the process by which light is transformed into chemical energy, to subsequently produce food in the form of sugars. Water (H_2O) and carbon dioxide (CO_2) are used in photosynthesis, and sugar and oxygen (O_2) is made, using light energy. Photosynthesis is divided into two stages—the light reactions, where water is split to produce oxygen, and the dark reactions, or Calvin cycle, which builds sugar molecules from carbon dioxide. The two phases are linked by the energy carriers adenosine triphosphate (ATP) and nicotinamide adenine dinucleotide phosphate ($NADP^+$).

Light Reactions

The light reactions of photosynthesis take place across the thylakoid membranes.

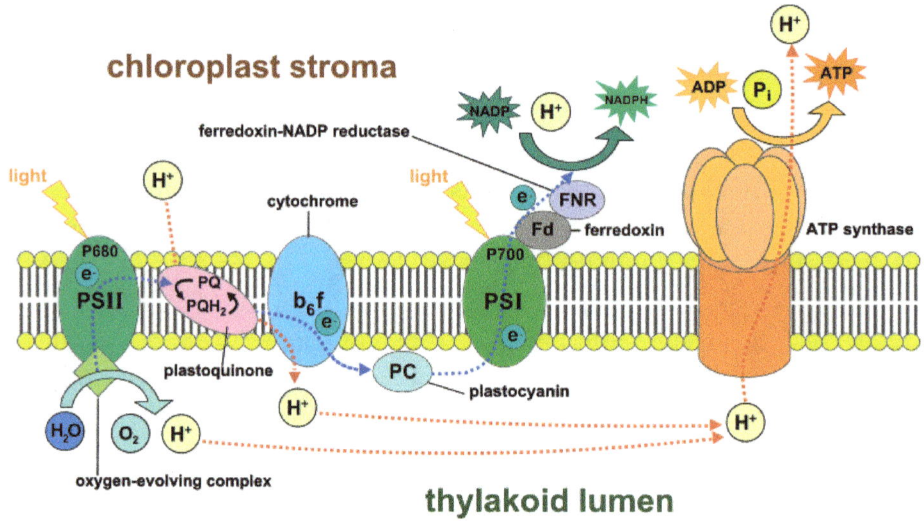

The light reactions take place on the thylakoid membranes. They take light energy and store it in NADPH, a form of $NADP^+$, and ATP to fuel the dark reactions.

Energy Carriers

ATP is the phosphorylated version of adenosine diphosphate (ADP), which stores energy in a cell and powers most cellular activities. ATP is the energized form, while ADP is the (partially) depleted form. $NADP^+$ is an electron carrier which ferries high energy electrons. In the light reactions, it gets reduced, meaning it picks up electrons, becoming NADPH.

Photophosphorylation

Like mitochondria, chloroplasts use the potential energy stored in an H^+, or hydrogen ion gradient to generate ATP energy. The two photosystems capture light energy to energize electrons taken from water, and release them down an electron transport chain. The molecules between the photosystems harness the electrons' energy to pump hydrogen ions into the thylakoid space, creating a concentration gradient, with more hydrogen ions (up to a thousand times as many) inside the thylakoid system than in the stroma. The hydrogen ions in the thylakoid space then diffuse back down their concentration gradient, flowing back out into the stroma through ATP synthase. ATP synthase uses the energy from the flowing hydrogen ions to phosphorylate adenosine diphosphate into adenosine triphosphate, or ATP. Because chloroplast ATP synthase projects out into the stroma, the ATP is synthesized there, in position to be used in the dark reactions.

$NADP^+$ Reduction

Electrons are often removed from the electron transport chains to charge $NADP^+$ with electrons, reducing it to NADPH. Like ATP synthase, ferredoxin-$NADP^+$reductase, the enzyme that reduces $NADP^+$, releases the NADPH it makes into the stroma, right where it is needed for the dark reactions.

Because $NADP^+$ reduction removes electrons from the electron transport chains, they must be replaced—the job of photosystem II, which splits water molecules (H_2O) to obtain the electrons from its hydrogen atoms.

Cyclic Photophosphorylation

While photosystem II photolyzes water to obtain and energize new electrons, photosystem I simply reenergizes depleted electrons at the end of an electron transport chain. Normally, the reenergized electrons are taken by $NADP^+$, though sometimes they can flow back down more H^+-pumping electron transport chains to transport more hydrogen ions into the thylakoid space to generate more ATP. This is termed cyclic photophosphorylation because the electrons are recycled. Cyclic photophosphorylation is common in C_4 plants, which need more ATP than NADPH.

Dark Reactions

The Calvin cycle, also known as the dark reactions, is a series of biochemical reactions that fixes CO_2 into G3P sugar molecules and uses the energy and electrons from the ATP and NADPH made in the light reactions. The Calvin cycle takes place in the stroma of the chloroplast.

While named *"the dark reactions"*, in most plants, they take place in the light, since the dark reactions are dependent on the products of the light reactions.

Carbon Fixation and G3P Synthesis

The Calvin cycle starts by using the enzyme Rubisco to fix CO_2 into five-carbon Ribulose bisphosphate (RuBP) molecules. The result is unstable six-carbon molecules that immediately break down into three-carbon molecules called 3-phosphoglyceric acid, or 3-PGA. The ATP and NADPH made in the light reactions is used to convert the 3-PGA into glyceraldehyde-3-phosphate, or G3P sugar molecules. Most of the G3P molecules are recycled back into RuBP using energy from more ATP, but one out of every six produced leaves the cycle—the end product of the dark reactions.

The Calvin cycle (Interactive diagram) The Calvin cycle incorporates carbon dioxide into sugar molecules.

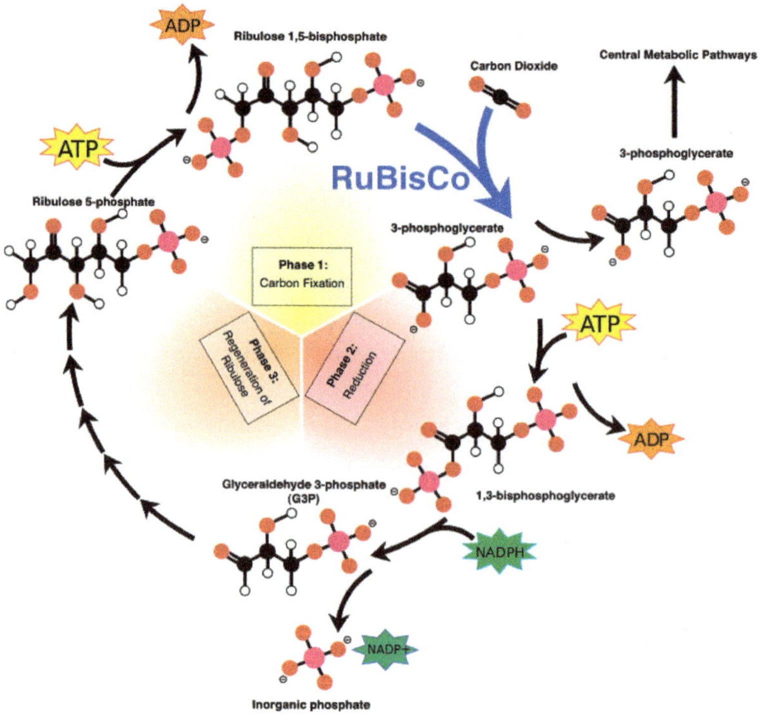

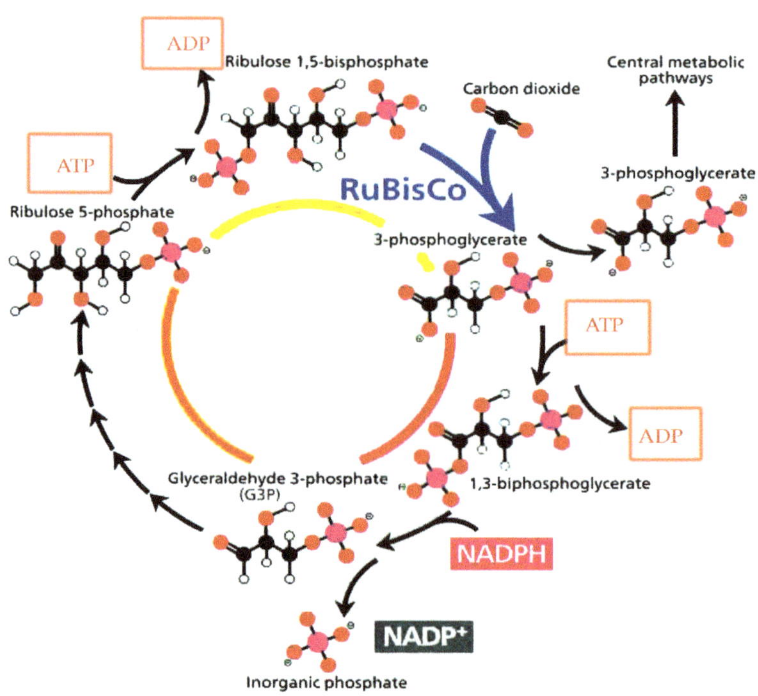

Sugars and Starches

Glyceraldehyde-3-phosphate can double up to form larger sugar molecules like glucose and fructose. These molecules are processed, and from them, the still larger sucrose, a disaccharide commonly known as table sugar, is made, though this process takes place outside of the chloroplast, in the cytoplasm.

Sucrose is made up of a glucose monomer (left), and a fructose monomer (right).

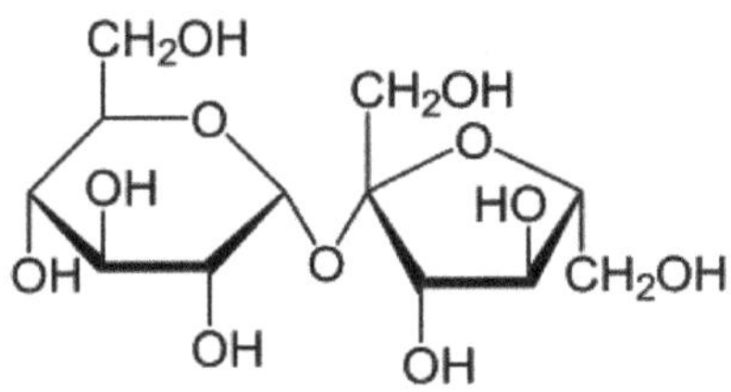

Alternatively, glucose monomers in the chloroplast can be linked together to make starch, which accumulates into the starch grains found in the chloroplast. Under conditions such as high atmospheric CO_2 concentrations, these starch grains may grow very large, distorting the grana and thylakoids. The starch granules displace the thylakoids, but leave them intact.Waterlogged roots can also cause starch buildup in the chloroplasts, possibly due to less sucrose being exported out of the chloroplast (or more accurately, the plant cell). This depletes a plant's free phosphate supply, which indirectly stimulates chloroplast starch synthesis. While linked to low photosynthesis rates, the starch grains themselves may not necessarily interfere significantly with the efficiency of photosynthesis, and might simply be a side effect of another photosynthesis-depressing factor.

Photorespiration

Photorespiration can occur when the oxygen concentration is too high. Rubisco cannot distinguish between oxygen and carbon dioxide very well, so it can accidentally add O_2 instead of CO_2 to RuBP. This process reduces the efficiency of photosynthesis—it consumes ATP and oxygen, releases CO_2, and produces no sugar. It can waste up to half the carbon fixed by the Calvin cycle. Several mechanisms have evolved in different lineages that raise the carbon dioxide concentration relative to oxygen within the chloroplast, increasing the efficiency of photosynthesis. These mechanisms are called carbon dioxide concentrating mechanisms, or CCMs. These include Crassulacean acid metabolism, C_4 carbon fixation, and pyrenoids. Chloroplasts in C_4 plants are notable as they exhibit a distinct chloroplast dimorphism.

pH

Because of the H^+ gradient across the thylakoid membrane, the interior of the thylakoid is acidic, with a pH around 4, while the stroma is slightly basic, with a pH of around 8. The optimal stroma pH for the Calvin cycle is 8.1, with the reaction nearly stopping when the pH falls below 7.3.

CO_2 in water can form carbonic acid, which can disturb the pH of isolated chloroplasts, interfering with photosynthesis, even though CO_2 is used in photosynthesis. However, chloroplasts in living plant cells are not affected by this as much.

Chloroplasts can pump K^+ and H^+ ions in and out of themselves using a poorly understood light-driven transport system.

In the presence of light, the pH of the thylakoid lumen can drop up to 1.5 pH units, while the pH of the stroma can rise by nearly one pH unit.

Amino Acid Synthesis

Chloroplasts alone make almost all of a plant cell's amino acids in their stroma except the sulfur-containing ones like cysteine and methionine. Cysteine is made in the chloroplast (the proplastid too) but it is also synthesized in the cytosol and mitochondria, probably because it has trouble crossing membranes to get to where it is needed. The chloroplast is known to make the precursors to methionine but it is unclear whether the organelle carries out the last leg of the pathway or if it happens in the cytosol.

Other Nitrogen Compounds

Chloroplasts make all of a cell's purines and pyrimidines—the nitrogenous bases found in DNA and RNA. They also convert nitrite (NO_2^-) into ammonia (NH_3) which supplies the plant with nitrogen to make its amino acids and nucleotides.

Other Chemical Products

Chloroplasts are the site of complex lipid metabolism.

Differentiation, Replication, and Inheritance

Chloroplasts are a special type of a plant cell organelle called a **plastid**, though the two terms are sometimes used interchangeably. There are many other types of plastids, which carry out various functions. All chloroplasts in a plant are descended from undifferentiated proplastids found in the zygote, or fertilized egg. Proplastids are commonly found in an adult plant's apical meristems. Chloroplasts do not normally develop from proplastids in root tip meristems—instead, the formation of starch-storing amyloplasts is more common.

In shoots, proplastids from shoot apical meristems can gradually develop into chloroplasts in photosynthetic leaf tissues as the leaf matures, if exposed to the required light. This process involves invaginations of the inner plastid membrane, forming sheets of membrane that project into the internal stroma. These membrane sheets then fold to form thylakoids and grana.

If angiosperm shoots are not exposed to the required light for chloroplast formation, proplastids may develop into an etioplast stage before becoming chloroplasts. An etioplast is a plastid that lacks chlorophyll, and has inner membrane invaginations that form a lattice of tubes in their stroma, called a prolamellar body. While etioplasts lack chlorophyll, they have a yellow chlorophyll precursor stocked. Within a few minutes of light exposure, the prolamellar body begins to reorganize into stacks of thylakoids, and chlorophyll starts to be produced. This process, where the etioplast becomes a chloroplast, takes several hours. Gymnosperms do not require light to form chloroplasts.

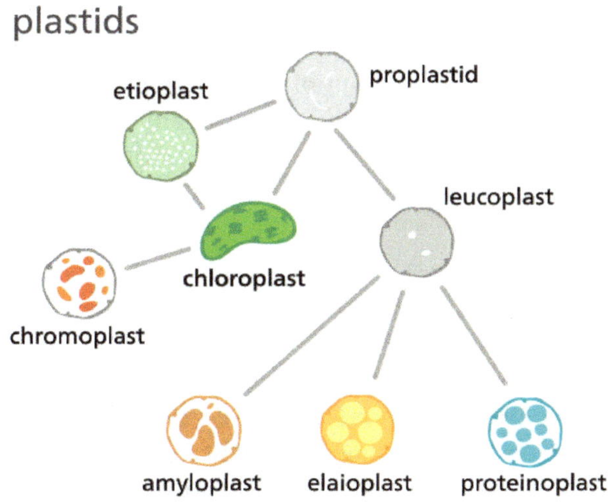

Plastid types *(Interactive diagram)* Plants contain many different kinds of plastids in their cells.

Light, however, does not guarantee that a proplastid will develop into a chloroplast. Whether a proplastid develops into a chloroplast some other kind of plastid is mostly controlled by the nucleus and is largely influenced by the kind of cell it resides in.

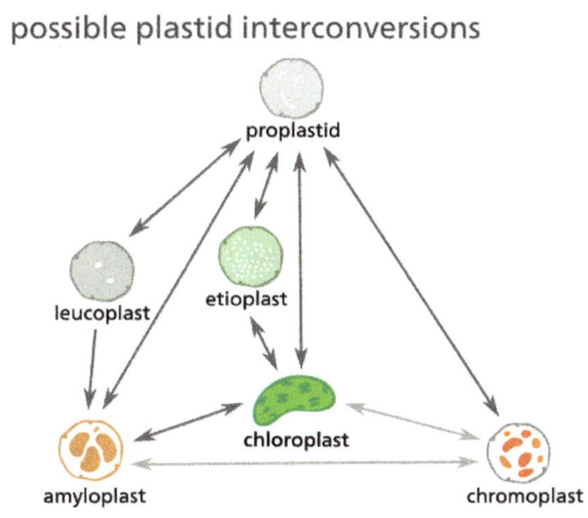

Many plastid interconversions are possible.

Plastid Interconversion

Plastid differentiation is not permanent, in fact many interconversions are possible. Chloroplasts may be converted to chromoplasts, which are pigment-filled plastids responsible for the bright colors seen in flowers and ripe fruit. Starch storing amyloplasts can also be converted to chromoplasts, and it is possible for proplastids to develop straight into chromoplasts. Chromoplasts and amyloplasts can also become chloroplasts, like what happens when a carrot or a potato is illuminated. If a plant is injured, or something else causes a plant cell to revert to a meristematic state, chloroplasts and other plastids can turn back into proplastids. Chloroplast, amyloplast, chromoplast, proplast, etc., are not absolute states—intermediate forms are common.

Chloroplast Division

Most chloroplasts in a photosynthetic cell do not develop directly from proplastids or etioplasts. In fact, a typical shoot meristematic plant cell contains only 7–20 proplastids. These proplastids differentiate into chloroplasts, which divide to create the 30–70 chloroplasts found in a mature photosynthetic plant cell. If the cell divides, chloroplast division provides the additional chloroplasts to partition between the two daughter cells.

In single-celled algae, chloroplast division is the only way new chloroplasts are formed. There is no proplastid differentiation—when an algal cell divides, its chloroplast divides along with it, and each daughter cell receives a mature chloroplast.

Almost all chloroplasts in a cell divide, rather than a small group of rapidly dividing chloroplasts. Chloroplasts have no definite S-phase—their DNA replication is not synchronized or limited to that of their host cells. Much of what we know about chloroplast division comes from studying organisms like *Arabidopsis* and the red alga *Cyanidioschyzon merolæ*.

The division process starts when the proteins FtsZ1 and FtsZ2 assemble into filaments, and with the help of a protein ARC6, form a structure called a Z-ring within the chloroplast's stroma. The Min system manages the placement of the Z-ring, ensuring that the chloroplast is cleaved more or less evenly. The protein MinD prevents FtsZ from linking up and forming filaments. Another protein ARC3 may also be involved, but it is not very well understood. These proteins are active at the poles of the chloroplast, preventing Z-ring formation there, but near the center of the chloroplast, MinE inhibits them, allowing the Z-ring to form.

Next, the two plastid-dividing rings, or PD rings form. The inner plastid-dividing ring is located in the inner side of the chloroplast›s inner membrane, and is formed first. The outer plastid-dividing ring is found wrapped around the outer chloroplast membrane. It consists of filaments about 5 nanometers across, arranged in rows 6.4 nanometers apart, and shrinks to squeeze the chloroplast. This is when chloroplast constriction begins.

In a few species like *Cyanidioschyzon merolæ*, chloroplasts have a third plastid-dividing ring located in the chloroplast's intermembrane space.

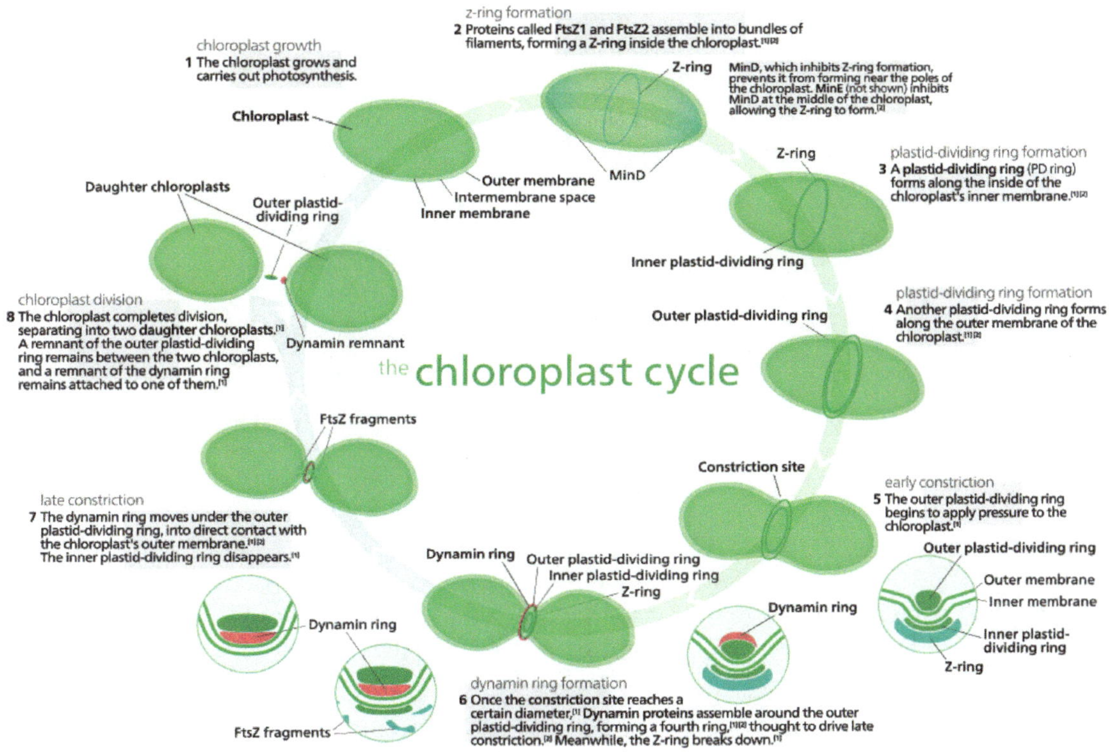

Most chloroplasts in plant cells, and all chloroplasts in algae arise from chloroplast division. *Picture references,*

Late into the constriction phase, dynamin proteins assemble around the outer plastid-dividing ring, helping provide force to squeeze the chloroplast.Meanwhile, the Z-ring and the inner plastid-dividing ring break down. During this stage, the many chloroplast DNA plasmids floating around in the stroma are partitioned and distributed to the two forming daughter chloroplasts.

Later, the dynamins migrate under the outer plastid dividing ring, into direct contact with the chloroplast's outer membrane, to cleave the chloroplast in two daughter chloroplasts.

A remnant of the outer plastid dividing ring remains floating between the two daughter chloroplasts, and a remnant of the dynamin ring remains attached to one of the daughter chloroplasts.

Of the five or six rings involved in chloroplast division, only the outer plastid-dividing ring is present for the entire constriction and division phase—while the Z-ring forms first, constriction does not begin until the outer plastid-dividing ring forms.

image · labels

Chloroplast division In this light micrograph of some moss chloroplasts, many dumbbell-shaped chloroplasts can be seen dividing. Grana are also just barely visible as small granules.

In species of algae that contain a single chloroplast, regulation of chloroplast division is extremely important to ensure that each daughter cell receives a chloroplast—chloroplasts can›t be made from scratch. In organisms like plants, whose cells contain multiple chloroplasts, coordination is looser and less important. It is likely that chloroplast and cell division are somewhat synchronized, though the mechanisms for it are mostly unknown.

Light has been shown to be a requirement for chloroplast division. Chloroplasts can grow and progress through some of the constriction stages under poor quality green light, but are slow to complete division—they require exposure to bright white light to complete division. Spinach leaves grown under green light have been observed to contain many large dumbbell-shaped chloroplasts. Exposure to white light can stimulate these chloroplasts to divide and reduce the population of dumbbell-shaped chloroplasts.

Chloroplast Inheritance

Like mitochondria, chloroplasts are usually inherited from a single parent. Biparental chloroplast inheritance—where plastid genes are inherited from both parent plants—occurs in very low levels in some flowering plants.

Many mechanisms prevent biparental chloroplast DNA inheritance, including selective destruction of chloroplasts or their genes within the gamete or zygote, and chloroplasts from one parent being excluded from the embryo. Parental chloroplasts can be sorted so that only one type is present in each offspring.

Gymnosperms, such as pine trees, mostly pass on chloroplasts paternally, while flowering plants often inherit chloroplasts maternally. Flowering plants were once thought to only inherit chloroplasts maternally. However, there are now many documented cases of angiosperms inheriting chloroplasts paternally.

Angiosperms, which pass on chloroplasts maternally, have many ways to prevent paternal inheritance. Most of them produce sperm cells that do not contain any plastids. There are many other documented mechanisms that prevent paternal inheritance in these flowering plants, such as different rates of chloroplast replication within the embryo.

Among angiosperms, paternal chloroplast inheritance is observed more often in hybrids than in offspring from parents of the same species. This suggests that incompatible hybrid genes might interfere with the mechanisms that prevent paternal inheritance.

Transplastomic Plants

Recently, chloroplasts have caught attention by developers of genetically modified crops. Since, in most flowering plants, chloroplasts are not inherited from the male parent, transgenes in these plastids cannot be disseminated by pollen. This makes plastid transformation a valuable tool for the creation and cultivation of genetically modified plants that are biologically contained, thus posing significantly lower environmental risks. This biological containment strategy is therefore suitable for establishing the coexistence of conventional and organic agriculture. While the reliability of this mechanism has not yet been studied for all relevant crop species, recent results in tobacco plants are promising, showing a failed containment rate of transplastomic plants at 3 in 1,000,000.

CHLOROPLAST DNA

Chloroplasts have their own DNA, often abbreviated as cpDNA. It is also known as the plastome when referring to genomes of other plastids. Its existence was first proved in 1962, and first sequenced in 1986—when two Japanese research teams sequenced the chloroplast DNA of liverwort and tobacco. Since then, hundreds of chloroplast DNAs from various species have been sequenced, but they are mostly those of land plants and green algae—glaucophytes, red algae, and other algal groups are extremely underrepresented, potentially introducing some bias in views of "typical" chloroplast DNA structure and content.

Linear Structure

Chloroplast DNAs have long been thought to have a circular structure, but some evidence suggests that chloroplast DNA more commonly takes a linear shape.Over 95% of the chloroplast DNA in corn chloroplasts has been observed to be in branched linear form rather than individual circles.

RNA Editing in Plastids

RNA editing is the insertion, deletion, and substitution of nucleotides in a mRNA transcript prior to translation to protein. The highly oxidative environment inside chloroplasts increases the rate of mutation so post-transcription repairs are needed to conserved functional sequences. The chloroplast editosome substitutes C -> U and U -> C at very specific locations on the transcript. This can change the codon for an amino acid or restore a non-functional pseudogene by adding an AUG start codon or removing a premature UAA stop codon.

The editosome recognizes and binds to cis sequence upstream of the editing site. The distance between the binding site and editing site varies by gene and proteins involved in the editosome. Hundreds of different PPR proteins from the nuclear genome are involved in the RNA editing process. These proteins consist of 35-mer repeated amino acids, the sequence of which determines the cis binding site for the edited transcript.

Basal land plants such as liverworts, mosses and ferns have hundreds of different editing sites while flowering plants typically have between thirty and forty. Parasitic plants such as Epifagus virginiana show a loss of RNA editing resulting in a loss of function for photosynthesis genes.

Cytoplasmic Translation and N-Terminal Transit Sequences

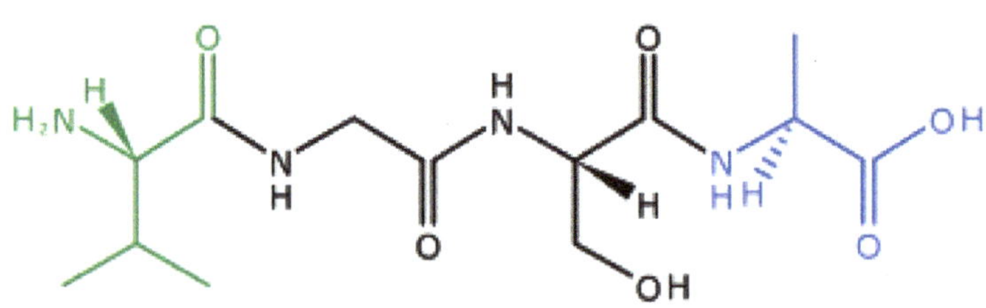

A polypeptide with four amino acids linked together. At the left is the N-terminus, with its amino (H_2N) group in green. The blue C-terminus, with its carboxyl group (CO_2H) is at the right.

Polypeptides, the precursors of proteins, are chains of amino acids. The two ends of a polypeptide are called the N-terminus, or *amino end*, and the C-terminus, or *carboxyl end*. For many (but not all)chloroplast proteins encoded by nuclear genes, *cleavable transit peptides* are added to the N-termini of the polypeptides, which are used to help direct the polypeptide to the chloroplast for import (N-terminal transit peptides are also used to direct polypeptides to plant mitochondria). N-terminal transit sequences are also called *presequences* because they are located at the "front" end of a polypeptide—ribosomes synthesize polypeptides from the N-terminus to the C-terminus.

Chloroplast transit peptides exhibit huge variation in length and amino acid sequence. They can be from 20–150 amino acids long—an unusually long length, suggesting that transit peptides are actually collections of domains with different functions. Transit peptides tend to be positively charged, rich in hydroxylated amino acids such as serine, threonine, and proline, and poor in acidic amino acids like aspartic acid and glutamic acid. In an aqueous solution, the transit sequence forms a random coil.

Not all chloroplast proteins include a N-terminal cleavable transit peptide though. Some include the transit sequence within the functional part of the protein itself. A few have their transit sequence appended to their C-terminus instead. Most of the polypeptides that lack N-terminal targeting sequences are the ones that are sent to the outer chloroplast membrane, plus at least one sent to the inner chloroplast membrane.

Phosphorylation, Chaperones, and Transport

After a chloroplast polypeptide is synthesized on a ribosome in the cytosol, ATP energy can be used to phosphorylate, or add a phosphate group to many (but not all) of them in their transit sequences. Serine and threonine (both very common in chloroplast transit sequences—making up 20–30% of the sequence) are often the amino acids that accept the phosphate group. The enzyme that carries out the phosphorylation is specific for chloroplast polypeptides, and ignores ones meant for mitochondria or peroxisomes.

Phosphorylation changes the polypeptide's shape, making it easier for 14-3-3 proteins to attach to the polypeptide. In plants, 14-3-3 proteins only bind to chloroplast preproteins. It is also bound by the **h**eat **s**hock **p**rotein Hsp70 that keeps the polypeptide from folding prematurely. This is important because it prevents chloroplast proteins from assuming their active form and carrying out their chloroplast functions in the wrong place—the cytosol. At the same time, they have to keep just enough shape so that they can be recognized and imported into the chloroplast.

The heat shock protein and the 14-3-3 proteins together form a cytosolic guidance complex that makes it easier for the chloroplast polypeptide to get imported into the chloroplast.

Alternatively, if a chloroplast preprotein's transit peptide is not phosphorylated, a chloroplast preprotein can still attach to a heat shock protein or Toc159. These complexes can bind to the TOC complex on the outer chloroplast membrane using GTP energy.

The Translocon on the Outer Chloroplast Membrane (TOC)

The TOC complex, or ***t****ranslocon on the* ***o****uter* ***c****hloroplast membrane*, is a collection of proteins that imports preproteins across the outer chloroplast envelope. Five subunits of the TOC complex have been identified—two GTP-binding proteins Toc34 and Toc159, the protein import tunnel Toc75, plus the proteins Toc64 and Toc12.

The first three proteins form a core complex that consists of one Toc159, four to five Toc34s, and four Toc75s that form four holes in a disk 13 nanometers across. The whole core complex weighs about 500 kilodaltons. The other two proteins, Toc64 and Toc12, are associated with the core complex but are not part of it.

Toc34 and 33

Toc34 is an integral protein in the outer chloroplast membrane that's anchored into it by its hydrophobicC-terminal tail. Most of the protein, however, including its large guanosine triphosphate (GTP)-binding domain projects out into the stroma.

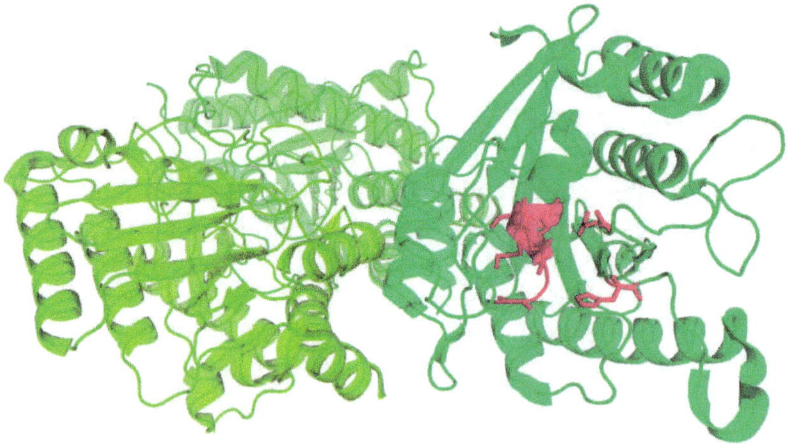

Toc34 from a pea plant. Toc34 has three almost identical molecules (shown in slightly different shades of green), each of which forms a dimer with one of its adjacent molecules. Part of a GDP molecule binding site is highlighted in pink.

Toc34's job is to catch some chloroplast preproteins in the cytosol and hand them off to the rest of the TOC complex. When GTP, an energy molecule similar to ATP attaches to Toc34, the protein becomes much more able to bind to many chloroplast preproteins in the cytosol. The chloroplast preprotein's presence causes Toc34 to break GTP into guanosine diphosphate (GDP) and inorganic phosphate. This loss of GTP makes the Toc34 protein release the chloroplast preprotein, handing it off to the next TOC protein. Toc34 then releases the depleted GDP molecule, probably with the help of an unknown GDP exchange factor. A domain of Toc159 might be the exchange factor that carry out the GDP removal. The Toc34 protein can then take up another molecule of GTP and begin the cycle again.

Toc34 can be turned off through phosphorylation. A protein kinase drifting around on the outer chloroplast membrane can use ATP to add a phosphate group to the Toc34 protein, preventing it from being able to receive another GTP molecule, inhibiting the protein's activity. This might provide a way to regulate protein import into chloroplasts.

Arabidopsis thaliana has two homologous proteins, AtToc33 and AtToc34 (The *At* stands for ***A**rabidopsis **t**haliana*), which are each about 60% identical in amino acid sequence to Toc34 in peas (called *ps*Toc34). AtToc33 is the most common in *Arabidopsis*, and it is the functional analogue of Toc34 because it can be turned off by phosphorylation. AtToc34 on the other hand cannot be phosphorylated.

Toc159

Toc159 is another GTP binding TOC subunit, like Toc34. Toc159 has three domains. At the N-terminal end is the A-domain, which is rich in acidic amino acids and takes up about half the protein length. The A-domain is often cleaved off, leaving an 86 kilodalton fragment called Toc86. In the middle is its GTP binding domain, which is very similar to the homologous GTP-binding domain in Toc34. At the C-terminal end is the hydrophilic M-domain, which anchors the protein to the outer chloroplast membrane.

Toc159 probably works a lot like Toc34, recognizing proteins in the cytosol using GTP. It can be regulated through phosphorylation, but by a different protein kinase than the one that phosphorylates Toc34. Its M-domain forms part of the tunnel that chloroplast preproteins travel through, and seems to provide the force that pushes preproteins through, using the energy from GTP. Toc159 is not always found as part of the TOC complex—it has also been found dissolved in the cytosol. This suggests that it might act as a shuttle that finds chloroplast preproteins in the cytosol and carries them back to the TOC complex. There isn't a lot of direct evidence for this behavior though.

A family of Toc159 proteins, Toc159, Toc132, Toc120, and Toc90 have been found in *Arabidopsis thaliana*. They vary in the length of their A-domains, which is completely gone in Toc90. Toc132, Toc120, and Toc90 seem to have specialized functions in importing stuff like nonphotosynthetic preproteins, and can't replace Toc159.

Toc75

β-barrel The general shape of a β-barrel is a hollow cylinder lined by multiple β-sheets. Note that the protein depicted is *not* Toc75 specifically.

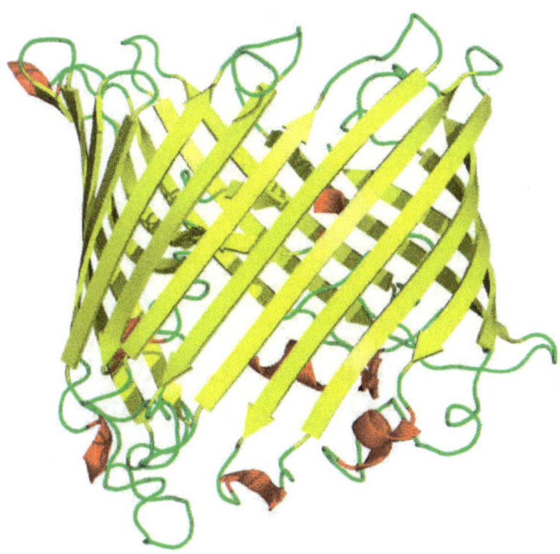

Toc75 is the most abundant protein on the outer chloroplast envelope. It is a transmembrane tube that forms most of the TOC pore itself. Toc75 is a β-barrel channel lined by 16 β-pleated sheets. The hole it forms is about 2.5 nanometers wide at the ends, and shrinks to about 1.4–1.6 nanometers in diameter at its narrowest point—wide enough to allow partially folded chloroplast preproteins to pass through.

Toc75 can also bind to chloroplast preproteins, but is a lot worse at this than Toc34 or Toc159.

Arabidopsis thaliana has multiple isoforms of Toc75 that are named by the chromosomal positions of the genes that code for them. AtToc75 III is the most abundant of these.

The Translocon on the Inner Chloroplast Membrane (TIC)

The TIC translocon, or ***t**ranslocon on the **i**nner **c**hloroplast membrane translocon* is another protein complex that imports proteins across the inner chloroplast envelope. Chloroplast polypeptide chains probably often travel through the two complexes at the same time, but the TIC complex can also retrieve preproteins lost in the intermembrane space.

Like the TOC translocon, the TIC translocon has a large core complex surrounded by some loosely associated peripheral proteins like Tic110, Tic40, and Tic21.The core complex weighs about one million daltons and contains Tic214, Tic100, Tic56, and Tic20 I, possibly three of each.

Tic20

Tic20 is an integral protein thought to have four transmembrane α-helices. It is found in the 1 million dalton TIC complex. Because it is similar to bacterialamino acid transporters and the mitochondrial import protein Tim17 (***t**ranslocase on the **i**nner **m**itochondrial **m**embrane*), it has been proposed to be part of the TIC import channel. There is no *in vitro* evidence for this though. In *Arabidopsis thaliana*, it is known that for about every five Toc75 proteins in the outer chloroplast membrane, there are two Tic20 I proteins (the main form of Tic20 in *Arabidopsis*) in the inner chloroplast membrane.

Unlike Tic214, Tic100, or Tic56, Tic20 has homologous relatives in cyanobacteria and nearly all chloroplast lineages, suggesting it evolved before the first chloroplast endosymbiosis. Tic214, Tic100, and Tic56 are unique to chloroplastidan chloroplasts, suggesting that they evolved later.

Tic214

Tic214 is another TIC core complex protein, named because it weighs just under 214 kilodaltons. It is 1786 amino acids long and is thought to have six transmembrane domains on its N-terminal end. Tic214 is notable for being coded for by chloroplast DNA, more specifically the first open reading frame *ycf1*. Tic214 and Tic20 together probably make up the part of the one million dalton TIC complex that spans the entire membrane. Tic20 is buried inside the complex while Tic214 is exposed on both sides of the inner chloroplast membrane.

Tic100

Tic100 is a nuclear encoded protein that's 871 amino acids long. The 871 amino acids collectively weigh slightly less than 100 thousand daltons, and since the mature protein probably doesn't lose any amino acids when itself imported into the chloroplast (it has no cleavable transit peptide), it was named Tic100. Tic100 is found at the edges of the

1 million dalton complex on the side that faces the chloroplast intermembrane space.

Tic56

Tic56 is also a nuclear encoded protein. The preprotein its gene encodes is 527 amino acids long, weighing close to 62 thousand daltons; the mature form probably undergoes processing that trims it down to something that weighs 56 thousand daltons when it gets imported into the chloroplast. Tic56 is largely embedded inside the 1 million dalton complex.

Tic56 and Tic100 are highly conserved among land plants, but they don't resemble any protein whose function is known. Neither has any transmembrane domains.

REFERENCES

1. Jones, Daniel (2003) , Peter Roach, James Hartmann and Jane Setter, eds., English Pronouncing Dictionary, Cambridge: Cambridge University Press, ISBN 3-12-539683-2
2. "Chloroplast". Merriam-Webster Dictionary.
3. "How large are chloroplasts?". Cell Biology By The Number. Ron Milo and Rob Philips.
4. "chloroplast". Online Etymology Dictionary.
5. von Mohl, H. 1835/1837. Ueber die Vermehrung der Pflanzen-Zellen durch Teilung. Dissert. Tubingen 1835. Flora 1837, .
6. Schimper, A.F.W. (1883). "Über die Entwicklung der Chlorophyllkörner und Farbkörper". Bot. Zeitung (in German). 41: 105–14, 121–31, 137–46, 153–62.
7. Strasburger, E. 1884. Das botanische Praktikum, 1. Aufl. Gustav Fischer: Jena
8. Gunning, B., Koenig, F. & Govindjee, P. (2006). A dedication to pioneers of research on chloroplast structure. In: R. R. Wise & J. K. Hoober (ed.). The structure and function of plastids. Springer: Netherlands, pp. xxiii-xxxi, .
9. Hoober, J. K. (1984). Chloroplasts. Plenum: New York, .
10. Biology 8th Edition Campbell & Reece. Benjamin Cummings (Pearson). 2009. p. 516.
11. Mereschkowsky C (1905). "Über Natur und Ursprung der Chromatophoren im Pflanzenreiche". Biol Centralbl (in German). 25: 593–604.
12. Alberts, Bruce (2002). Molecular biology of the cell (4. ed.). New York u.a.: Garland. ISBN 0-8153-4072-9.
13. Kumar, K.; Mella-Herrera, R. A.; Golden, J. W. (2010). "Cyanobacterial Heterocysts". Cold Spring Harbor Perspectives in Biology. 2 (4): a000315. doi:10.1101/cshperspect.a000315. PMC 2845205 . PMID 20452939.
14. Biology 8th edition—Campbell&Reece. Benjamin Cummings. 2008. pp. 186–187. ISBN 978-0-321-54325-7.

15. Kim, E.; Archibald, J. M. (2009). "Diversity and Evolution of Plastids and Their Genomes". In Sandelius, Anna Stina; Aronsson, Henrik. The Chloroplast. Plant Cell Monographs. 13. pp. 1–39. doi:10.1007/978-3-540-68696-5_1. ISBN 978-3-540-68692-7.

16. Bryant, Donald A.; Guglielmi, Gérard; Marsac, Nicole Tandeau; Castets, Anne-Marie; Cohen-Bazire, Germaine (1979). "The structure of cyanobacterial phycobilisomes: A model". Archives of Microbiology. 123 (2): 311–34. doi:10.1007/BF00446810.

17. Keeling, Patrick J. (2004). "Diversity and evolutionary history of plastids and their hosts". American Journal of Botany. 91 (10): 1481–93. doi:10.3732/ajb.91.10.1481. PMID 21652304.

18. McFadden, Geoffrey I; Van Dooren, Giel G (2004). "Evolution: Red Algal Genome Affirms a Common Origin of All Plastids". Current Biology. 14 (13): R514–6. doi:10.1016/j.cub.2004.06.041. PMID 15242632.

19. Fuks, Bruno; Fabrice Homblé (8 July 1996). "Mechanism of Proton Permeation through Chloroplast Lipid Membranes" (PDF). Plant Physiology. 112 (2): 759–766. doi:10.1104/pp.112.2.759. PMC 158000 . PMID 8883387. Retrieved 25 August 2013.

20. Joyard J Block MA, Douce R; Block; Douce (1991). "Molecular aspects of plastid envelope biochemistry". Eur J Biochem. 199 (3): 489–509. doi:10.1111/j.1432-1033.1991.tb16148.x. PMID 1868841.

21. "Chloroplast". Encyclopedia of Science. Retrieved 27 December 2012.

22. Nakayama, Takuro; Archibald, John M (2012). "Evolving a photosynthetic organelle". BMC Biology. 10 (1): 35. doi:10.1186/1741-7007-10-35. PMC 3337241 . PMID 22531210.

23. McFadden, G. I. (2001). "Chloroplast Origin and Integration". Plant Physiology. 125 (1): 50–3. doi:10.1104/pp.125.1.50. PMC 1539323 . PMID 11154294.

24. Archibald, John M. (2009). "The Puzzle of Plastid Evolution". Current Biology. 19 (2): R81–8. doi:10.1016/j.cub.2008.11.067. PMID 19174147.

25. Delaye, Luis; Valadez-Cano, Cecilio; Pérez-Zamorano, Bernardo (2016). "How Really Ancient Is Paulinella Chromatophora?". PLoS Currents. doi:10.1371/currents.tol.e68a099364bb1a1e129a17b4e06b0c6b.

26. Wise, edited by Robert R.; Hoober, J. Kenneth (2006). The structure and function of plastids. Dordrecht: Springer. pp. 3–21. ISBN 978-1-4020-4061-0.

27. Ball, S.; Colleoni, C.; Cenci, U.; Raj, J. N.; Tirtiaux, C. (2011). "The evolution of glycogen and starch metabolism in eukaryotes gives molecular clues to understand the establishment of plastid endosymbiosis". Journal of Experimental Botany. 62 (6): 1775–801. doi:10.1093/jxb/erq411. PMID 21220783.

28. Nair, Sethu C.; Striepen, Boris (2011). "What Do Human Parasites Do with a Chloroplast Anyway?". PLoS Biology. 9 (8): e1001137. doi:10.1371/journal.pbio.1001137. PMC 3166169 . PMID 21912515.

29. Keeling, P. J. (2010). "The endosymbiotic origin, diversification and fate of plastids". Philosophical Transactions of the Royal Society B: Biological Sciences. 365 (1541): 729–48. doi:10.1098/rstb.2009.0103. PMC 2817223 . PMID 20124341.

30. Biology 8th Edition Campbell & Reece. Benjamin Cummings (Pearson). 2009. pp. 582–92.

31. "rhodo-". The Free Dictionary. Farlex. Retrieved 2013-06-07.

32. Lewis, L. A.; McCourt, R. M. (2004). "Green algae and the origin of land plants". American Journal of Botany. 91 (10): 1535–56. doi:10.3732/ajb.91.10.1535. PMID 21652308.

33. Machida, M.; Takechi, K; Sato, H; Chung, SJ; Kuroiwa, H; Takio, S; Seki, M; Shinozaki, K; Fujita, T; Hasebe, M; Takano, H (2006). "Genes for the peptidoglycan synthesis pathway are essential for chloroplast division in moss". Proceedings of the National Academy of Sciences. 103 (17): 6753–8. Bibcode:2006PNAS..103.6753M. doi:10.1073/pnas.0510693103. PMC 1458953 . PMID 16618924.

34. Moroney, J. V.; Somanchi, A (1999). "How Do Algae Concentrate CO2 to Increase the Efficiency of Photosynthetic Carbon Fixation?". Plant Physiology. 119 (1): 9–16. doi:10.1104/pp.119.1.9. PMC 1539202 . PMID 9880340.

35. Tartar, AuréLien; Boucias, Drion G. (2004). "The non-photosynthetic, pathogenic green alga Helicosporidium sp. Has retained a modified, functional plastid genome". FEMS Microbiology Letters. 233 (1): 153–7. doi:10.1016/j.femsle.2004.02.006. PMID 15043882.

36. Chaal, Balbir K.; Green, Beverley R. (February 2005). "Protein import pathways in 'complex' chloroplasts derived from secondary endosymbiosis involving a red algal ancestor". Plant Molecular Biology. 57 (3): 333–42. doi:10.1007/s11103-004-7848-y. PMID 15830125.

37. Moustafa, Ahmed; Beszteri, Bánk; Maier, Uwe G.; Bowler, Chris; Valentin, Klaus; Bhattacharya, Debashish (2009). "Genomic Footprints of a Cryptic Plastid Endosymbiosis in Diatoms". Science. 324 (5935): 1724–6. Bibcode:2009Sci...324.1724M. doi:10.1126/science.1172983. PMID 19556510.

38. Rogers, M. B.; Gilson, P. R.; Su, V.; McFadden, G. I.; Keeling, P. J. (2006). "The Complete Chloroplast Genome of the Chlorarachniophyte Bigelowiella natans: Evidence for Independent Origins of Chlorarachniophyte and Euglenid Secondary Endosymbionts". Molecular Biology and Evolution. 24 (1): 54–62. doi:10.1093/molbev/msl129. PMID 16990439.

39. Hackett, J. D.; Anderson, D. M.; Erdner, D. L.; Bhattacharya, D. (2004). "Dinoflagellates: A remarkable evolutionary experiment". American Journal of Botany. 91 (10): 1523–34. doi:10.3732/ajb.91.10.1523. PMID 21652307.

40. Dorrell, R. G.; Smith, A. G. (2011). "Do Red and Green Make Brown?: Perspectives on Plastid Acquisitions within Chromalveolates". Eukaryotic Cell. 10 (7): 856–68. doi:10.1128/EC.00326-10. PMC 3147421 . PMID 21622904.

41. Meeson, B. W.; Chang, S. S.; Sweeney, B. M. (1982). "Characterization of

Peridinin-Chlorophyll α-Proteins from the Marine Dinoflagellate Ceratium furca". Botanica Marina. 25 (8): 347–50. doi:10.1515/botm.1982.25.8.347.

42. Tengs, T; Dahlberg, OJ; Shalchian-Tabrizi, K; Klaveness, D; Rudi, K; Delwiche, CF; Jakobsen, KS (2000). "Phylogenetic analyses indicate that the 19'Hexanoyloxy-fucoxanthin-containing dinoflagellates have tertiary plastids of haptophyte origin". Molecular Biology and Evolution. 17 (5): 718–29. doi:10.1093/oxfordjournals.molbev.a026350. PMID 10779532.
43. Schnepf, Eberhard; Elbrächter, Malte (1999). "Dinophyte chloroplasts and phylogeny – A review". Grana. 38 (2–3): 81–97. doi:10.1080/00173139908559217.
44. Nowack, E. C. M.; Vogel, H.; Groth, M.; Grossman, A. R.; Melkonian, M.; Glöckner, G. (2010). "Endosymbiotic Gene Transfer and Transcriptional Regulation of Transferred Genes in Paulinella chromatophora". Molecular Biology and Evolution. 28 (1): 407–22. doi:10.1093/molbev/msq209. PMID 20702568.
45. Dann, Leighton (2002). Bioscience—Explained (PDF). Green DNA: BIOSCIENCE EXPLAINED.
46. Clegg, M. T.; Gaut, BS; Learn Jr, GH; Morton, BR (1994). "Rates and Patterns of Chloroplast DNA Evolution". Proceedings of the National Academy of Sciences. 91 (15): 6795–801. Bibcode:1994PNAS...91.6795C. doi:10.1073/pnas.91.15.6795. PMC 44285 . PMID 8041699.
47. Shaw, J.; Lickey, E. B.; Schilling, E. E.; Small, R. L. (2007). "Comparison of whole chloroplast genome sequences to choose noncoding regions for phylogenetic studies in angiosperms: The tortoise and the hare III". American Journal of Botany. 94 (3): 275–88. doi:10.3732/ajb.94.3.275. PMID 21636401.
48. Skovgaard, Alf (1998). "Role of chloroplast retention in a marine dinoflagellate". Aquatic Microbial Ecology. 15: 293–301. doi:10.3354/ame015293.
49. Hogan, C. Michael (2010). Deoxyribonucleic acid. Encyclopedia of Earth. National Council for Science and the Environment. eds. S.Draggan and C.Cleveland. Washington DC
50. "ctDNA — chloroplast DNA". AllAcronyms.com. Retrieved 2 January 2013.
51. The Oxford Dictionary of Abbreviations. ctDNA—Dictionary definition. 1998.
52. "Chloroplasts and Other Plastids". University of Hamburg. Retrieved 27 December 2012.
53. Sandelius, Anna Stina (2009). The Chloroplast: Interactions with the Environment. Springer. p. 18. ISBN 978-3-540-68696-5.
54. Burgess, Jeremy (1989). An introduction to plant cell development. Cambridge: Cambridge university press. p. 62. ISBN 0-521-31611-1.
55. Bendich, A. J. (2004). "Circular Chloroplast Chromosomes: The Grand Illusion". The Plant Cell Online. 16 (7): 1661–6. doi:10.1105/tpc.160771. PMC 514151 . PMID 15235123.
56. Kolodner, R.; Tewari, KK (1979). "Inverted Repeats in Chloroplast DNA from

Higher Plants". Proceedings of the National Academy of Sciences. 76 (1): 41–5. Bibcode:1979PNAS...76...41K. doi:10.1073/pnas.76.1.41. PMC 382872 . PMID 16592612.

57. Palmer, Jeffrey D.; Thompson, William F. (1982). "Chloroplast DNA rearrangements are more frequent when a large inverted repeat sequence is lost". Cell. 29 (2): 537–50. doi:10.1016/0092-8674(82)90170-2. PMID 6288261.
58. Plant Biochemistry (3rd ed.). Academic Press. 2005. p. 517. ISBN 978-0-12-088391-2.
59. Kobayashi, T.; Takahara, M; Miyagishima, SY; Kuroiwa, H; Sasaki, N; Ohta, N; Matsuzaki, M; Kuroiwa, T (2002). "Detection and Localization of a Chloroplast-Encoded HU-Like Protein That Organizes Chloroplast Nucleoids". The Plant Cell Online. 14 (7): 1579–89. doi:10.1105/tpc.002717. PMC 150708 . PMID 12119376.
60. Krishnan, Rao BJ, NM (2009). "A comparative approach to elucidate chloroplast genome replication". BMC Genomics. 10 (237). doi:10.1186/1471-2164-10-237.
61. Heinhorst, Gordon C. Cannon, Sabine (1993). "DNA replication in chloroplasts". Journal of Cell Science. 104: 1–9.
62. Bendich, AJ (2004). "Circular Chloroplast Chromosomes: The Grand Illusion". The Plant Cell. 16 (7): 1661–1666. doi:10.1105/tpc.160771. PMC 514151 . PMID 15235123.
63. "Effect of chemical mutagens on nucleotide sequence". Biocyclopedia. Retrieved 24 October 2015.
64. Alberts, Bruce (2002). Molecular biology of the cell (4. ed.). New York u.a. Garland. ISBN 0-8153-4072-9.
65. Huang, Chun Y.; Ayliffe, Michael A.; Timmis, Jeremy N. (2003). "Direct measurement of the transfer rate of chloroplast DNA into the nucleus". Nature. 422 (6927): 72–6. Bibcode:2003Natur.422...72H. doi:10.1038/nature01435. PMID 12594458.
66. Martin, W.; Rujan, T; Richly, E; Hansen, A; Cornelsen, S; Lins, T; Leister, D; Stoebe, B; Hasegawa, M; Penny, D (2002). "Evolutionary analysis of Arabidopsis, cyanobacterial, and chloroplast genomes reveals plastid phylogeny and thousands of cyanobacterial genes in the nucleus". Proceedings of the National Academy of Sciences. 99 (19): 12246–51. Bibcode:2002PNAS...9912246M. doi:10.1073/pnas.182432999. PMC 129430 . PMID 12218172.
67. Smith, David Roy; Lee, Robert W. (2014-04-01). "A Plastid without a Genome: Evidence from the Nonphotosynthetic Green Algal Genus Polytomella". Plant Physiology. 164 (4): 1812–1819. doi:10.1104/pp.113.233718. PMC 3982744 . PMID 24563281.
68. Archibald, John M. (2006). "Algal Genomics: Exploring the Imprint of Endosymbiosis". Current Biology. 16 (24): R1033–5. doi:10.1016/j.cub.2006.11.008. PMID 17174910.
69. Koussevitzky, Shai; Nott, A; Mockler, TC; Hong, F; Sachetto-Martins, G;

Surpin, M; Lim, J; Mittler, R; Chory, J (2007). "Signals from Chloroplasts Converge to Regulate Nuclear Gene Expression". Science. 316 (5825): 715–9. Bibcode:2007Sci...316..715K. doi:10.1126/science.1140516 (inactive 2015-01-01). PMID 17395793.

70. Hedtke, B.; Börner, T; Weihe, A (1997). "Mitochondrial and Chloroplast Phage-Type RNA Polymerases in Arabidopsis". Science. 277 (5327): 809–11. doi:10.1126/science.277.5327.809. PMID 9242608.
71. Harris, EH; Boynton, JE; Gillham, NW (1994). "Chloroplast ribosomes and protein synthesis". Microbiological reviews. 58 (4): 700–54. PMC 372988 . PMID 7854253.
72. Soll, Jürgen; Schleiff, Enrico (2004). "Plant cell biology: Protein import into chloroplasts". Nature Reviews Molecular Cell Biology. 5 (3): 198–208. doi:10.1038/nrm1333. PMID 14991000.
73. Biology 8th edition—Campbell & Reece. Benjamin Cummings. 2008. p. 340. ISBN 978-0-321-54325-7.
74. Lung, S.-C.; Chuong, S. D. X. (2012). "A Transit Peptide-Like Sorting Signal at the C Terminus Directs the Bienertia sinuspersici Preprotein Receptor Toc159 to the Chloroplast Outer Membrane". The Plant Cell. 24 (4): 1560–78. doi:10.1105/tpc.112.096248. PMC 3398564 . PMID 22517318.
75. Soll, J r.; Soll, J (1996). "Phosphorylation of the Transit Sequence of Chloroplast Precursor Proteins". Journal of Biological Chemistry. 271 (11): 6545–54. doi:10.1074/jbc.271.11.6545. PMID 8626459.
76. May, T. (2000). "14-3-3 Proteins Form a Guidance Complex with Chloroplast Precursor Proteins in Plants". The Plant Cell Online. 12: 53–64. doi:10.1105/tpc.12.1.53.
77. Jarvis, Paul; Soll, Jürgen (2001). "Toc, Tic, and chloroplast protein import". Biochimica et Biophysica Acta (BBA) - Molecular Cell Research. 1541 (1–2): 64–79. doi:10.1016/S0167-4889(01)00147-1. PMID 11750663.
78. Wise, R.R.; Hoober, J.K. (2007). The Structure and Function of Plastids. Springer. pp. 32–33. ISBN 978-1-4020-6570-5.
79. "Oedogonium Link ex Hirn, 1900: 17". algaeBASE. Retrieved 19 May 2013.
80. "Chlamydomonas Ehrenberg, 1833: 288". algaeBASE. Retrieved 19 May 2013.
81. "Spirogyra Link, 1820: 5". algaeBASE. Retrieved 19 May 2013.
82. "Sirogonium Kützing, 1843: 278". algaeBASE. Retrieved 19 May 2013.
83. "Zygnema C.Agardh, 1817: xxxii, 98". algaeBASE. Retrieved 19 May 2013.
84. "Micrasterias C.Agardh ex Ralfs, 1848: 68". algaeBASE. Retrieved 19 May 2013.
85. John, D.M.; Brook, A.J.; Whitton, B.A. (2002). The freshwater algal flora of the British Isles : an identification guide to freshwater and terrestrial algae. Cambridge: Cambridge University Press. p. 335. ISBN 978-0-521-77051-4.
86. Biology 8th edition—Campbell & Reece. Benjamin Cummings. 2008. pp. 196–197. ISBN 978-0-321-54325-7.

87. Koike, H.; Yoshio, M.; Kashino, Y.; Satoh, K. (1998). "Polypeptide Composition of Envelopes of Spinach Chloroplasts: Two Major Proteins Occupy 90% of Outer Envelope Membranes". Plant and Cell Physiology. 39 (5): 526–32. doi:10.1093/oxfordjournals.pcp.a029400. PMID 9664716.
88. Köhler RH, Hanson MR (1 January 2000). "Plastid tubules of higher plants are tissue-specific and developmentally regulated". J. Cell. Sci. 113 (Pt 1): 81–9. PMID 10591627. Archived from the original on 2010-11-18.
89. Gray JC, Sullivan JA, Hibberd JM, Hansen MR (2001). "Stromules: mobile protrusions and interconnections between plastids". Plant Biology. 3 (3): 223–33. doi:10.1055/s-2001-15204.
90. Schattat, M.; Barton, K.; Baudisch, B.; Klösgen, R. B.; Mathur, J. (2011). "Plastid Stromule Branching Coincides with Contiguous Endoplasmic Reticulum Dynamics". Plant Physiology. 155 (4): 1667–77. doi:10.1104/pp.110.170480. PMC 3091094 . PMID 21273446.
91. Schattat, M. H.; Griffiths, S.; Mathur, N.; Barton, K.; Wozny, M. R.; Dunn, N.; Greenwood, J. S.; Mathur, J. (2012). "Differential Coloring Reveals That Plastids Do Not Form Networks for Exchanging Macromolecules". The Plant Cell. 24 (4): 1465–77. doi:10.1105/tpc.111.095398. PMC 3398557 . PMID 22474180.
92. Brunkard, Jacob O.; Runkel, Anne M.; Zambryski, Patricia C. (2015). "Chloroplasts extend stromules independently and in response to internal redox signals". PNAS. 112 (32): 10044–10049. doi:10.1073/pnas.1511570112.
93. Burgess,, Jeremy (1989). An introduction to plant cell development (Pbk. ed.). Cambridge: Cambridge university press. p. 46. ISBN 0-521-31611-1.
94. Whatley, Jean M (5 July 1994). "The occurrence of a peripheral reticulum in plastids of the gymnosperm Welwitschia mirabilis". New Phytologist. 74 (2): 215–220. doi:10.1111/j.1469-8137.1975.tb02608.x.
95. Wise, Robert R (2007). The Structure and Function of Plastids. Springer. pp. 17–18. ISBN 978-1-4020-6570-5.
96. Manuell, Andrea L.; Quispe, Joel; Mayfield, Stephen P. (2007). "Structure of the Chloroplast Ribosome: Novel Domains for Translation Regulation". PLoS Biology. 5 (8): e209. doi:10.1371/journal.pbio.0050209. PMC 1939882 . PMID 17683199.
97. Lim, K.; Kobayashi, I.; Nakai, K. (7 April 2014). "Alterations in rRNA-mRNA Interaction during Plastid Evolution". Molecular Biology and Evolution. 31 (7): 1728–1740. doi:10.1093/molbev/msu120.
98. Hirose, T; Sugiura, M (January 2004). "Functional Shine-Dalgarno-like sequences for translational initiation of chloroplast mRNAs.". Plant & cell physiology. 45 (1): 114–7. doi:10.1093/pcp/pch002. PMID 14749493.
99. Ma, J; Campbell, A; Karlin, S (October 2002). "Correlations between Shine-Dalgarno sequences and gene features such as predicted expression levels and operon structures.". Journal of Bacteriology. 184 (20): 5733–45. doi:10.1128/JB.184.20.5733-5745.2002. PMID 12270832.

100. Lim, K; Furuta, Y; Kobayashi, I (October 2012). "Large variations in bacterial ribosomal RNA genes.". Molecular Biology and Evolution. 29 (10): 2937–48. doi:10.1093/molbev/mss101. PMID 22446745.

101. Austin, J. R.; Frost, E; Vidi, PA; Kessler, F; Staehelin, LA (2006). "Plastoglobules Are Lipoprotein Subcompartments of the Chloroplast That Are Permanently Coupled to Thylakoid Membranes and Contain Biosynthetic Enzymes". The Plant Cell Online. 18 (7): 1693–703. doi:10.1105/tpc.105.039859. PMC 1488921 . PMID 16731586.

102. Crumpton-Taylor, M.; Grandison, S.; Png, K. M. Y.; Bushby, A. J.; Smith, A. M. (2011). "Control of Starch Granule Numbers in Arabidopsis Chloroplasts". Plant Physiology. 158 (2): 905–16. doi:10.1104/pp.111.186957. PMC 3271777 . PMID 22135430.

103. Zeeman, Samuel C.; Delatte, Thierry; Messerli, Gaëlle; Umhang, Martin; Stettler, Michaela; Mettler, Tabea; Streb, Sebastian; Reinhold, Heike; Kötting, Oliver (2007). "Starch breakdown: Recent discoveries suggest distinct pathways and novel mechanisms". Functional Plant Biology. 34 (6): 465–73. doi:10.1071/FP06313.

104. J D, Rochaix (1998). The molecular biology of chloroplasts and mitochondria in Chlamydomonas. Dordrecht u.a.: Kluwer Acad. Publ. pp. 550–565. ISBN 978-0-7923-5174-0.

105. Steer, Brian E.S. Gunning, Martin W. (1996). Plant cell biology : structure and function. Boston, Mass.: Jones and Bartlett Publishers. p. 24. ISBN 0-86720-504-0.

106. Hanson, David; Andrews, T. John; Badger, Murray R. (2002). "Variability of the pyrenoid-based CO2 concentrating mechanism in hornworts (Anthocerotophyta)". Functional Plant Biology. 29 (3): 407–16. doi:10.1071/PP01210.

107. Ma, Y.; Pollock, S. V.; Xiao, Y.; Cunnusamy, K.; Moroney, J. V. (2011). "Identification of a Novel Gene, CIA6, Required for Normal Pyrenoid Formation in Chlamydomonas reinhardtii". Plant Physiology. 156 (2): 884–96. doi:10.1104/pp.111.173922. PMC 3177283 . PMID 21527423.

108. Retallack, B; Butler, RD (1970). "The development and structure of pyrenoids in Bulbochaete hiloensis". Journal of Cell Science. 6 (1): 229–41. PMID 5417694.

109. Brown, Malcolm R; Howard J Arnott (1970). "Structure and Function of the Algal Pyrenoid" (PDF). Journal of Phycology. 6: 14–22. doi:10.1111/j.1529-8817.1970.tb02350.x. Retrieved 31 December 2012.

110. Mustardy, L.; Buttle, K.; Steinbach, G.; Garab, G. (2008). "The Three-Dimensional Network of the Thylakoid Membranes in Plants: Quasihelical Model of the Granum-Stroma Assembly". The Plant Cell Online. 20 (10): 2552–7. doi:10.1105/tpc.108.059147. PMC 2590735 . PMID 18952780.

111. Austin, J. R.; Staehelin, L. A. (2011). "Three-Dimensional Architecture of Grana and Stroma Thylakoids of Higher Plants as Determined by Electron Tomography". Plant Physiology. 155 (4): 1601–11. doi:10.1104/pp.110.170647. PMC 3091084 . PMID 21224341.

112. Infanger, S; Bischof, S; Hiltbrunner, A; Agne, B; Baginsky, S; Kessler, F (2011).

"The Chloroplast Import Receptor Toc90 Partially Restores the Accumulation of Toc159 Client Proteins in the Arabidopsis thaliana ppi2 Mutant". Molecular Plant. 4 (2): 252–63. doi:10.1093/mp/ssq071. PMID 21220583.

113. "thylakoid". Merriam-Webster Dictionary. Merriam-Webster. Retrieved 19 May 2013.
114. Biology 8th Edition Campbell & Reece. Benjamin Cummings (Pearson). 2009. pp. 190–193.
115. "Australian scientists discover first new chlorophyll in 60 years". University of Sydney. 20 August 2010.
116. Takaichi, Shinichi (15 June 2011). "Carotenoids in Algae: Distributions, Biosyntheses and Functions". Marine Drugs. 9 (12): 1101–1118. doi:10.3390/md9061101. PMC 3131562 . PMID 21747749.
117. Shapley, Dan. "Why Do Leaves Change Color in Fall?". News Articles. Retrieved 21 May 2013.
118. Howe, C.J; Barbrook, A.C; Nisbet, R.E.R; Lockhart, P.J; Larkum, A.W.D (2008). "The origin of plastids". Philosophical Transactions of the Royal Society B: Biological Sciences. 363 (1504): 2675–85. doi:10.1098/rstb.2008.0050. PMC 2606771 . PMID 18468982.
119. "Introduction to the Rhodophyta". University of California Museum of Paleontology. Retrieved 20 May 2013.
120. Biology 8th Edition Campbell & Reece. Benjamin Cummings (Pearson). 2009. pp. 200–201.
121. Lawton, June R (March 1988). "Ultrastructure of Chloroplast Membranes in Leaves of Maize and Ryegrass as Revealed by Selective Staining Methods". New Phytologist. 108 (3): 277–283. doi:10.1111/j.1469-8137.1988.tb04163.x. JSTOR 2433294.
122. Roberts, editor, Keith (2007). Handbook of plant science. Chichester, West Sussex, England: Wiley. p. 16. ISBN 978-0-470-05723-0.
123. Biology 8th edition—Campbell&Reece. Benjamin Cummings. 2008. p. 742. ISBN 978-0-321-54325-7.
124. Lawson T. and J. I. L. Morison. Essay 10.1 Guard Cell Photosynthesis. Plant Physiology and Development, Sixth Edition
125. Eckardt, N. A. (2003). "Controlling Organelle Positioning: A Novel Chloroplast Movement Protein". The Plant Cell Online. 15 (12): 2755–7. doi:10.1105/tpc.151210. PMC 540263 .
126. Glynn, Jonathan M.; Miyagishima, Shin-ya; Yoder, David W.; Osteryoung, Katherine W.; Vitha, Stanislav (2007). "Chloroplast Division". Traffic. 8 (5): 451–61. doi:10.1111/j.1600-0854.2007.00545.x. PMID 17451550.
127. Dong, X.-J.; Nagai, R.; Takagi, S. (1998). "Microfilaments Anchor Chloroplasts along the Outer Periclinal Wall in Vallisneria Epidermal Cells through Cooperation of PFR and Photosynthesis". Plant and Cell Physiology. 39 (12): 1299–306. doi:10.1093/oxfordjournals.pcp.a029334.

128. Takagi, Shingo (December 2002). "Actin-based photo-orientation movement of chloroplasts in plant cells". Journal of Experimental Biology. 206: 1963–1969. doi:10.1242/jeb.00215. Retrieved 6 March 2013.
129. Zeiger, E.; Talbott, L. D.; Frechilla, S.; Srivastava, A.; Zhu, J. (2002). "The guard cell chloroplast: A perspective for the twenty-first century". New Phytologist. 153 (3): 415–424. doi:10.1046/j.0028-646X.2001.NPH328.doc.x.
130. Padmanabhan, Meenu S.; Dinesh-Kumar, S. P. (2010). "All Hands on Deck—The Role of Chloroplasts, Endoplasmic Reticulum, and the Nucleus in Driving Plant Innate Immunity". Molecular Plant-Microbe Interactions. 23 (11): 1368–80. doi:10.1094/MPMI-05-10-0113. PMID 20923348.
131. Katsir, Leron; Chung, Hoo Sun; Koo, Abraham JK; Howe, Gregg A (2008). "Jasmonate signaling: A conserved mechanism of hormone sensing". Current Opinion in Plant Biology. 11 (4): 428–35. doi:10.1016/j.pbi.2008.05.004. PMC 2560989 . PMID 18583180.
132. Schnurr, J. A.; Shockey, JM; De Boer, GJ; Browse, JA (2002). "Fatty Acid Export from the Chloroplast. Molecular Characterization of a Major Plastidial Acyl-Coenzyme a Synthetase from Arabidopsis". Plant Physiology. 129 (4): 1700–9. doi:10.1104/pp.003251. PMC 166758 . PMID 12177483.
133. Biology—Concepts and Connections. Pearson. 2009. pp. 108–118.
134. Campbell, Neil A.; Williamson, Brad; Heyden, Robin J. (2006). Biology: Exploring Life. Boston, Massachusetts: Pearson Prentice Hall. ISBN 978-0-13-250882-7.
135. Jagendorf A. T. and E. Uribe (1966). "ATP formation caused by acid-base transition of spinach chloroplasts.". Proc. Natl. Acad. Sci. USA. 55: 170–177. doi:10.1073/pnas.55.1.170.
136. Clarke, Jeremy M. Berg; John L. Tymoczko; Lubert Stryer. Web content by Neil D. (2002). Biochemistry (5. ed., 4. print. ed.). New York, NY u.a.: W. H. Freeman. pp. Section 19.4. ISBN 0-7167-3051-0.
137. Clarke, Jeremy M. Berg; John L. Tymoczko; Lubert Stryer. Web content by Neil D. (2002). Biochemistry (5. ed., 4. print. ed.). New York, NY u.a.: W. H. Freeman. pp. Section 20.1. ISBN 0-7167-3051-0.
138. Wample, R. L.; Davis, R. W. (1983). "Effect of Flooding on Starch Accumulation in Chloroplasts of Sunflower (Helianthus annuus L.)". Plant Physiology. 73 (1): 195–8. doi:10.1104/pp.73.1.195. PMC 1066435 . PMID 16663176.
139. Carmi, A.; Shomer, I. (1979). "Starch Accumulation and Photosynthetic Activity in Primary Leaves of Bean (Phaseolus vulgaris L.)". Annals of Botany. 44 (4): 479–484.
140. Berg JM; Tymoczko JL; Stryer L. (2002). Biochemistry. (5th ed.). W H Freeman. pp. Section 19.4. Retrieved 30 October 2012.
141. Hauser, M; Eichelmann, H; Oja, V; Heber, U; Laisk, A (1995). "Stimulation by Light of Rapid pH Regulation in the Chloroplast Stroma in Vivo as Indicated by CO2 Solubilization in Leaves". Plant Physiology. 108 (3): 1059–1066. doi:10.1104/pp.108.3.1059. PMC 157457 . PMID 12228527.

142. Werdan, Karl; Heldt, Hans W.; Milovancev, Mirjana (1975). "The role of pH in the regulation of carbon fixation in the chloroplast stroma. Studies on CO2 fixation in the light and dark". Biochimica et Biophysica Acta (BBA) - Bioenergetics. 396 (2): 276–92. doi:10.1016/0005-2728(75)90041-9. PMID 239746.

143. Burgess,, Jeremy (1989). An introduction to plant cell development. Cambridge: Cambridge university press. p. 56. ISBN 0-521-31611-1.

144. Ferro, Myriam; Daniel Salvi; Hélène Rivière-Rolland; Thierry Vermat; Daphné Seigneurin-Berny; Didier Grunwald; Jérôme Garin; Jacques Joyard; Norbert Rolland (20 August 2002). "Integral membrane proteins of the chloroplast envelope: Identification and subcellular localization of new transporters". Proceedings of the National Academy of Sciences. 99 (17): 11487–11492. Bibcode:2002PNAS...9911487F. doi:10.1073/pnas.172390399. PMC 123283 . PMID 12177442.

145. Rolland, Norbert; Michel Droux; Roland Douce (March 1992). "Subcellular Distribution of O-Acetylserine(thiol)lyase in Cauliflower (Brassica oleracea L.) Inflorescence". Plant Physiology. 98 (3): 927–935. doi:10.1104/pp.98.3.927.

146. Ravanel, Stéphane; Bertrand Gakière; Dominique Job; Roland Douce (23 June 1998). "The specific features of methionine biosynthesis and metabolism in plants". Proceedings of the National Academy of Sciences. 95 (13): 7805–7812. Bibcode:1998PNAS...95.7805R. doi:10.1073/pnas.95.13.7805. PMID 9636232.

147. Joyard, Jacques; Ferro, Myriam; Masselon, Christophe; Seigneurin-Berny, Daphné; Salvi, Daniel; Garin, Jérôme; Rolland, Norbert (2010). "Chloroplast proteomics highlights the subcellular compartmentation of lipid metabolism". Progress in Lipid Research. 49 (2): 128–58. doi:10.1016/j.plipres.2009.10.003. PMID 19879895.

148. Steer, Brian E.S. Gunning, Martin W. (1996). Plant cell biology : structure and function. Boston, Mass.: Jones and Bartlett Publishers. p. 20. ISBN 0-86720-504-0.

149. Burgess,, Jeremy (1989). An introduction to plant cell development(Pbk. ed.). Cambridge: Cambridge university press. pp. 54–55. ISBN 0-521-31611-1.

150. Burgess,, Jeremy (1989). An introduction to plant cell development (Pbk. ed.). Cambridge: Cambridge university press. p. 57. ISBN 0-521-31611-1.

151. Possingham, J. V.; Rose, R. J. (1976). "Chloroplast Replication and Chloroplast DNA Synthesis in Spinach Leaves". Proceedings of the Royal Society B: Biological Sciences. 193 (1112): 295–305. Bibcode:1976RSPSB.193..295P. doi:10.1098/rspb.1976.0047.

152. Heinhorst, S; G.C. Cannon (January 1993). "DNA replication in chloroplasts". Journal of Cell Science. 104 (104): 1–9. Retrieved 3 February 2013.

153. Miyagishima, S.-y.; Nishida, K; Mori, T; Matsuzaki, M; Higashiyama, T; Kuroiwa, H; Kuroiwa, T (2003). "A Plant-Specific Dynamin-Related Protein Forms a Ring at the Chloroplast Division Site". The Plant Cell Online. 15 (3): 655–65. doi:10.1105/tpc.009373. PMC 150020 . PMID 12615939.

154. Hashimoto, H.; Possingham, J. V. (1989). "Effect of Light on the Chloroplast Division Cycle and DNA Synthesis in Cultured Leaf Discs of Spinach". Plant

Physiology. 89 (4): 1178–83. doi:10.1104/pp.89.4.1178. PMC 1055993 . PMID 16666681.

155. Hansen, A. K.; Escobar, L. K.; Gilbert, L. E.; Jansen, R. K. (2007). "Paternal, maternal, and biparental inheritance of the chloroplast genome in Passiflora (Passifloraceae): Implications for phylogenetic studies". American Journal of Botany. 94 (1): 42–6. doi:10.3732/ajb.94.1.42. PMID 21642206.
156. Birky, C. William (1995). "Uniparental Inheritance of Mitochondrial and Chloroplast Genes: Mechanisms and Evolution". Proceedings of the National Academy of Sciences. 92 (25): 11331–8. Bibcode:1995PNAS...9211331B. doi:10.1073/pnas.92.25.11331. PMC 40394 . PMID 8524780.
157. Powell W, Morgante M, McDevitt R, Vendramin GG, Rafalski JA (August 1995). "Polymorphic simple sequence repeat regions in chloroplast genomes: applications to the population genetics of pines". Proc. Natl. Acad. Sci. U.S.A. 92 (17): 7759–63. Bibcode:1995PNAS...92.7759P. doi:10.1073/pnas.92.17.7759. PMC 41225 . PMID 7644491. In the pines, the chloroplast genome is transmitted through pollen
158. Stegemann, S.; Hartmann, S; Ruf, S; Bock, R (2003). "High-frequency gene transfer from the chloroplast genome to the nucleus". Proceedings of the National Academy of Sciences. 100 (15): 8828–33. Bibcode:2003PNAS..100.8828S. doi:10.1073/pnas.1430924100. PMC 166398 . PMID 12817081.
159. Ruf, S.; Karcher, D.; Bock, R. (2007). "Determining the transgene containment level provided by chloroplast transformation". Proceedings of the National Academy of Sciences. 104 (17): 6998–7002. Bibcode:2007PNAS..104.6998R. doi:10.1073/pnas.0700008104. PMC 1849964 . PMID 17420459.
160. C.Michael Hogan. 2010. Deoxyribonucleic acid. Encyclopedia of Earth. National Council for Science and the Environment. eds. S.Draggan and C.Cleveland. Washington DC
161. The Oxford Dictionary of Abbreviations. ctDNA—Dictionary definition. 1998.
162. Dann, Leighton (2002). Bioscience—Explained (PDF). Green DNA: BIOSCIENCE EXPLAINED.
163. "Chloroplasts and Other Plastids". University of Hamburg. Retrieved 27 December 2012.
164. Sandelius, Anna Stina (2009). The Chloroplast: Interactions with the Environment. Springer. p. 18. ISBN 978-3-540-68696-5.
165. Clegg, M. T.; Gaut, BS; Learn Jr, GH; Morton, BR (1994). "Rates and Patterns of Chloroplast DNA Evolution". Proceedings of the National Academy of Sciences. 91 (15): 6795–801. Bibcode:1994PNAS...91.6795C. doi:10.1073/pnas.91.15.6795. PMC 44285 . PMID 8041699.
166. Shaw, J.; Lickey, E. B.; Schilling, E. E.; Small, R. L. (2007). "Comparison of whole chloroplast genome sequences to choose noncoding regions for phylogenetic studies in angiosperms: The tortoise and the hare III". American Journal of Botany. 94 (3): 275–88. doi:10.3732/ajb.94.3.275. PMID 21636401.

167. Burgess, Jeremy (1989). An introduction to plant cell development. Cambridge: Cambridge university press. p. 62. ISBN 0-521-31611-1.

168. McFadden, G. I. (2001). "Chloroplast Origin and Integration". Plant Physiology. 125 (1): 50–3. doi:10.1104/pp.125.1.50. PMC 1539323 . PMID 11154294.

169. Kolodner, R.; Tewari, KK (1979). "Inverted Repeats in Chloroplast DNA from Higher Plants". Proceedings of the National Academy of Sciences. 76 (1): 41–5. Bibcode:1979PNAS...76...41K. doi:10.1073/pnas.76.1.41. PMC 382872 . PMID 16592612.

170. Palmer, Jeffrey D.; Thompson, William F. (1982). "Chloroplast DNA rearrangements are more frequent when a large inverted repeat sequence is lost". Cell. 29 (2): 537–50. doi:10.1016/0092-8674(82)90170-2. PMID 6288261.

171. Bendich, A. J. (2004). "Circular Chloroplast Chromosomes: The Grand Illusion". The Plant Cell Online. 16 (7): 1661–6. doi:10.1105/tpc.160771. PMC 514151 . PMID 15235123.

172. Plant Biochemistry (3rd ed.). Academic Press. 2005. p. 517. ISBN 9780120883912.

173. Biology 8th Edition Campbell & Reece. Benjamin Cummings (Pearson). 2009. p. 516.

174. Kobayashi, T.; Takahara, M; Miyagishima, SY; Kuroiwa, H; Sasaki, N; Ohta, N; Matsuzaki, M; Kuroiwa, T (2002). "Detection and Localization of a Chloroplast-Encoded HU-Like Protein That Organizes Chloroplast Nucleoids". The Plant Cell Online. 14 (7): 1579–89. doi:10.1105/tpc.002717. PMC 150708 . PMID 12119376.

175. Krishnan, Rao BJ, NM (2009). "A comparative approach to elucidate chloroplast genome replication". BMC Genomics. 10 (237): 237. doi:10.1186/1471-2164-10-237.

176. Heinhorst, Gordon C. Cannon, Sabine (1993). "DNA replication in chloroplasts". Journal of Cell Science. 104: 1–9.

177. Bendich, AJ (2004). "Circular Chloroplast Chromosomes: The Grand Illusion". The Plant Cell. 16 (7): 1661–1666. doi:10.1105/tpc.160771. PMC 514151 . PMID 15235123.

178. "Effect of chemical mutagens on nucleotide sequence". Biocyclopedia. Retrieved 24 October 2015.

179. Alberts, Bruce (2002). Molecular biology of the cell (4. ed.). New York u.a.: Garland. ISBN 0-8153-4072-9.

180. Harris, EH; Boynton, JE; Gillham, NW (1994). "Chloroplast ribosomes and protein synthesis". Microbiological reviews. 58 (4): 700–54. PMC 372988 . PMID 7854253.

181. Wakasugi, T.; Sugita, M.; Tsudzuki, T.; Sugiura, M. (1998). "Updated gene map of tobacco chloroplast DNA". Plant Molecular Biology Reporter. 16 (3): 231–41. doi:10.1023/A:1007564209282.

182. Krause, Kirsten (2008). "From chloroplasts to "cryptic" plastids: Evolution of plastid genomes in parasitic plants". Current Genetics. 54 (3): 111–21.

doi:10.1007/s00294-008-0208-8. PMID 18696071.

183. Peng, L.; Fukao, Y.; Fujiwara, M.; Shikanai, T. (2012). "Multistep Assembly of Chloroplast NADH Dehydrogenase-Like Subcomplex a Requires Several Nucleus-Encoded Proteins, Including CRR41 and CRR42, in Arabidopsis". The Plant Cell. 24 (1): 202–14. doi:10.1105/tpc.111.090597. PMC 3289569 . PMID 22274627.

184. Huang, Chun Y.; Ayliffe, Michael A.; Timmis, Jeremy N. (2003). "Direct measurement of the transfer rate of chloroplast DNA into the nucleus". Nature. 422 (6927): 72–6. Bibcode:2003Natur.422...72H. doi:10.1038/nature01435. PMID 12594458.

185. Martin, W.; Rujan, T; Richly, E; Hansen, A; Cornelsen, S; Lins, T; Leister, D; Stoebe, B; Hasegawa, M; Penny, D (2002). "Evolutionary analysis of Arabidopsis, cyanobacterial, and chloroplast genomes reveals plastid phylogeny and thousands of cyanobacterial genes in the nucleus". Proceedings of the National Academy of Sciences. 99 (19): 12246–51. Bibcode:2002PNAS...9912246M. doi:10.1073/pnas.182432999. PMC 129430 . PMID 12218172.

186. Mackiewicz, Paweł; Bodył, Andrzej; Moszczyński, Krzysztof (2014). "The case of horizontal gene transfer from bacteria to the peculiar dinoflagellate plastid genome". Mobile Genetic Elements. 3 (4): e25845. doi:10.4161/mge.25845. ISSN 2159-256X.

187. Leliaert, Frederik; Lopez-Bautista, Juan M (2015). "The chloroplast genomes of Bryopsis plumosa and Tydemania expeditionis (Bryopsidales, Chlorophyta): compact genomes and genes of bacterial origin". BMC Genomics. 16 (1): 204. doi:10.1186/s12864-015-1418-3. ISSN 1471-2164.

188. Moustafa, Ahmed; Beszteri, Bánk; Maier, Uwe G.; Bowler, Chris; Valentin, Klaus; Bhattacharya, Debashish (2009). "Genomic Footprints of a Cryptic Plastid Endosymbiosis in Diatoms". Science. 324 (5935): 1724–6. Bibcode:2009Sci...324.1724M. doi:10.1126/science.1172983. PMID 19556510.

189. Nowack, E. C. M.; Vogel, H.; Groth, M.; Grossman, A. R.; Melkonian, M.; Glöckner, G. (2010). "Endosymbiotic Gene Transfer and Transcriptional Regulation of Transferred Genes in Paulinella chromatophora". Molecular Biology and Evolution. 28 (1): 407–22. doi:10.1093/molbev/msq209. PMID 20702568.

190. Archibald, John M. (2006). "Algal Genomics: Exploring the Imprint of Endosymbiosis". Current Biology. 16 (24): R1033–5. doi:10.1016/j.cub.2006.11.008. PMID 17174910.

191. Koussevitzky, Shai; Nott, A; Mockler, TC; Hong, F; Sachetto-Martins, G; Surpin, M; Lim, J; Mittler, R; Chory, J (2007). "Signals from Chloroplasts Converge to Regulate Nuclear Gene Expression". Science. 316 (5825): 715–9. Bibcode:2007Sci...316..715K. doi:10.1126/science.1140516 (inactive 2015-01-11). PMID 17395793.

192. Hedtke, B.; Börner, T; Weihe, A (1997). "Mitochondrial and Chloroplast Phage-Type RNA Polymerases in Arabidopsis". Science. 277 (5327): 809–11. doi:10.1126/science.277.5327.809. PMID 9242608.

193. Takenaka, Mizuki; Zehrmann, Anja; Verbitskiy, Daniil; Härtel, Barbara; Brennicke, Axel (2013). "RNA Editing in Plants and Its Evolution". Annual Review of Genetics. 47 (1): 335–352. doi:10.1146/annurev-genet-111212-133519. ISSN 0066-4197.

194. Tillich, Michael; Krause, Kirsten (2010-07-31). "The ins and outs of editing and splicing of plastid RNAs: lessons from parasitic plants". New Biotechnology. Special Issue: Biotechnology Annual Review 2010RNA Basics and Biotechnology Applications. 27 (3): 256–266. doi:10.1016/j.nbt.2010.02.020.

195. Soll, Jürgen; Schleiff, Enrico (2004). "Plant cell biology: Protein import into chloroplasts". Nature Reviews Molecular Cell Biology. 5 (3): 198–208. doi:10.1038/nrm1333. PMID 14991000.

196. Keeling, P. J. (2010). "The endosymbiotic origin, diversification and fate of plastids". Philosophical Transactions of the Royal Society B: Biological Sciences. 365 (1541): 729–48. doi:10.1098/rstb.2009.0103. PMC 2817223 . PMID 20124341.

197. Biology 8th edition—Campbell & Reece. Benjamin Cummings. 2008. p. 340. ISBN 978-0-321-54325-7.

198. Wise, Robert R.; Hoober, J. Kenneth (2007). Structure and function of plastids. Berlin: Springer. pp. 53–74. ISBN 978-1-4020-6570-5.

199. Lee, D. W.; Lee, S; Lee, GJ; Lee, KH; Kim, S; Cheong, GW; Hwang, I (2006). "Functional Characterization of Sequence Motifs in the Transit Peptide of Arabidopsis Small Subunit of Rubisco". Plant Physiology. 140 (2): 466–83. doi:10.1104/pp.105.074575. PMC 1361317 . PMID 16384899.

200. May, T. (2000). "14-3-3 Proteins Form a Guidance Complex with Chloroplast Precursor Proteins in Plants". The Plant Cell Online. 12: 53–64. doi:10.1105/tpc.12.1.53.

201. Lung, S.-C.; Chuong, S. D. X. (2012). "A Transit Peptide-Like Sorting Signal at the C Terminus Directs the Bienertia sinuspersici Preprotein Receptor Toc159 to the Chloroplast Outer Membrane". The Plant Cell. 24 (4): 1560–78. doi:10.1105/tpc.112.096248. PMC 3398564 . PMID 22517318.

202. Soll, J r.; Soll, J (1996). "Phosphorylation of the Transit Sequence of Chloroplast Precursor Proteins". Journal of Biological Chemistry. 271 (11): 6545–54. doi:10.1074/jbc.271.11.6545. PMID 8626459.

203. Jarvis, Paul; Soll, Jürgen (2001). "Toc, Tic, and chloroplast protein import". Biochimica et Biophysica Acta (BBA) - Molecular Cell Research. 1541 (1–2): 64–79. doi:10.1016/S0167-4889(01)00147-1. PMID 11750663.

204. Sun, Yuh-Ju; Forouhar, Farhad; Li Hm, Hsou-min; Tu, Shuh-Long; Yeh, Yi-Hong; Kao, Sen; Shr, Hui-Lin; Chou, Chia-Cheng; Chen, Chinpan; Hsiao, Chwan-Deng (2001). "Crystal structure of pea Toc34, a novel GTPase of the chloroplast protein translocon". Nature Structural Biology. 9 (2): 95–100. doi:10.1038/nsb744. PMID 11753431.

205. Agne, B.; Andrès, C.; Montandon, C.; Christ, B.; Ertan, A.; Jung, F.; Infanger, S.; Bischof, S.; Baginsky, S.; Kessler, F. (2010). "The Acidic A-Domain of

Arabidopsis Toc159 Occurs as a Hyperphosphorylated Protein". Plant Physiology. 153 (3): 1016–30. doi:10.1104/pp.110.158048. PMC 2899928 . PMID 20457805.

206. Kikuchi, Shingo; Bédard, Jocelyn; Hirano, Minako; Hirabayashi, Yoshino; Oishi, Maya; Imai, Midori; Takase, Mai; Ide, Toru; Nakai, Masato (2013). "Uncovering the Protein Translocon at the Chloroplast Inner Envelope Membrane". Science. 339 (6119): 571–4. Bibcode:2013Sci...339..571K. doi:10.1126/science.1229262. PMID 23372012.

207. Sean P Curran and Carla M Koehler (2004). Mitochondrial Function and Biogenesis. Springer. p. 59. ISBN 9783540214892.

CHAPTER 4

Light-Dependent Reactions

Chapter Contents

The overall function of light-dependent reactions, the first stage of photosynthesis, is to convert solar energy into chemical energy in the form of NADPH and ATP, which are used in light-independent reactions and fuel the assembly of sugar molecules.

INTRODUCTION

In photosynthesis, the light-dependent reactions take place on the thylakoid membranes. The inside of the thylakoid membrane is called the lumen, and outside the thylakoid membrane is the stroma, where the light-independent reactions take place. The thylakoid membrane contains some integral membrane protein complexes that catalyze the light reactions. There are four major protein complexes in the thylakoid membrane: Photosystem II (PSII), Cytochrome b6f complex, Photosystem I (PSI), and ATP synthase. These four complexes work together to ultimately create the products ATP and NADPH.

The two photosystems absorb light energy through pigments - primarily the chlorophylls, which are responsible for the green color of leaves. The light-dependent reactions begin in photosystem II. When a chlorophyll *a* molecule within the reaction center of PSII absorbs a photon, an electron in this molecule attains a higher energy level. Because this state of an electron is very unstable, the electron is transferred from one to another molecule creating a chain of redox reactions, called an electron transport chain (ETC). The electron flow goes from PSII to cytochrome b6f to PSI. In PSI, the electron gets the energy from another photon. The final electron acceptor is NADP. In oxygenic photosynthesis, the first electron donor is water, creating oxygen as a waste product. In anoxygenic photosynthesis various electron donors are used.

Cytochrome b6f and ATP synthase work together to create ATP. This process is called photophosphorylation, which occurs in two different ways. In non-cyclic photophosphorylation, cytochrome b6f uses the energy of electrons from PSII to pump protons from the stroma to the lumen. The proton gradient across the thylakoid membrane creates a proton-motive force, used by ATP synthase to form ATP. In cyclic photophosphorylation, cytochrome b6f uses the energy of electrons from not only PSII but also PSI to create more ATP and to stop the production of NADPH. Cyclic phosphorylation is important to create ATP and maintain NADPH in the right proportion for the light-independent reactions.

The net-reaction of all light-dependent reactions in oxygenic photosynthesis is:

$$2H_2O + 2NADP^+ + 3ADP + 3P_i \rightarrow O_2 + 2NADPH + 3ATP$$

The two photosystems are protein complexes that absorb photons and are able to use this energy to create an electron transport chain. Photosystem I and II are very similar in structure and function. They use special proteins, called light-harvesting complexes, to absorb the photons with very high effectiveness. If a special pigment molecule in a photosynthetic reaction center absorbs a photon, an electron in this pigment attains the excited state and then is transferred to another molecule in the reaction center. This reaction, called photoinduced charge separation, is the start of the electron flow and is unique because it transforms light energy into chemical forms.

THE REACTION CENTER

The reaction center is in the thylakoid membrane. It transfers light energy to a dimer of chlorophyll pigment molecules near the periplasmic (or thylakoid lumen) side of the membrane. This dimer is called a special pair because of its fundamental role in photosynthesis. This special pair is slightly different in PSI and PSII reaction center. In PSII, it absorbs photons with a wavelength of 680 nm, and it is therefore called P680. In PSI, it absorbs photons at 700 nm, and it is called P700. In bacteria, the special pair is called P760, P840, P870, or P960.

If an electron of the special pair in the reaction center becomes excited, it cannot transfer this energy to another pigment using resonance energy transfer. In normal circumstances, the electron should return to the ground state, but, because the reaction center is arranged so that a suitable electron acceptor is nearby, the excited electron can move from the initial molecule to the acceptor. This process results in the formation of a positive charge on the special pair (due to the loss of an electron) and a negative charge on the acceptor and is, hence, referred to as photoinduced charge separation. In other words, electrons in pigment molecules can exist at specific energy levels. Under normal circumstances, they exist at the lowest possible energy level they can. However, if there is enough energy to move them into the next energy level, they can absorb that energy and occupy that higher energy level. The light they absorb contains the necessary amount of energy needed to push them into the next level. Any light that does not have enough or has too much energy cannot be absorbed and is reflected. The electron in the higher energy level, however, does not want to be there; the electron is unstable and must return to its normal lower energy level. To do this, it must release the energy that has put it into the higher energy state to begin with. This can happen various ways. The extra energy can be converted into molecular motion and lost as heat. Some of the extra energy can be lost as heat energy, while the rest is lost as light. (This re-emission of light energy is called fluorescence.) The energy, but not the e- itself, can be passed onto another molecule. (This is called resonance.) The energy and the e- can be transferred to another molecule. Plant pigments usually utilize the last two of these reactions to convert the sun's energy into their own.

This initial charge separation occurs in less than 10 picoseconds (10^{-11} seconds). In their high-energy states, the special pigment and the acceptor could undergo charge recombination; that is, the electron on the acceptor could move back to neutralize the positive charge on the special pair. Its return to the special pair would waste a valuable high-energy electron and simply convert the absorbed light energy into heat. In the case of PSII, this backflow of electrons can produce reactive oxygen species leading to photoinhibition. Three factors in the structure of the reaction center work together to suppress charge recombination nearly completely.

Another electron acceptor is less than 10 Å away from the first acceptor, and so the electron is rapidly transferred farther away from the special pair.

- An electron donor is less than 10 Å away from the special pair, and so the positive charge is neutralized by the transfer of another electron
- The electron transfer back from the electron acceptor to the positively charged special pair is especially slow. The rate of an of electron transfer reaction increases with its thermodynamic favorability up to a point and then decreases. The back transfer is so favourable that it takes place in the inverted region where electron-transfer rates become slower.

Thus, electron transfer proceeds efficiently from the first electron acceptor to the next, creating an electron transport chain that ends if it has reached NADPH.

Photosynthetic Electron Transport Chains in Chloroplasts

The photosynthesis process in chloroplasts begins when an electron of P680 of PSII attains a higher-energy level. This energy is used to reduce a chain of electron acceptors that have subsequently lowered redox-potentials. This chain of electron acceptors is known as an electron transport chain. When this chain reaches PS I, an electron is again excited, creating a high redox-potential. The electron transport chain of photosynthesis is often put in a diagram called the z-scheme, because the redox diagram from P680 to P700 resembles the letter z.

The final product of PSII is plastoquinol, a mobile electron carrier in the membrane. Plastoquinol transfers the electron from PSII to the proton pump, cytochrome b6f. The ultimate electron donor of PSII is water. Cytochrome b6f proceeds the electron chain to PSI through plastocyanin molecules. PSI is able to continue the electron transfer in two different ways. It can transfer the electrons either to plastoquinol again, creating a cyclic electron flow, or to an enzyme called FNR (Ferredoxin—NADP(+) reductase), creating a non-cyclic electron flow. PSI releases FNR into the stroma, where it reduces $NADP^+$ to NADPH.

Activities of the electron transport chain, especially from cytochrome b6f, lead to pumping of protons from the stroma to the lumen. The resulting transmembrane proton gradient is used to make ATP via ATP synthase.

The overall process of the photosynthetic electron transport chain in chloroplasts is:

$$\mathbf{H_2O \rightarrow PS\ II \rightarrow plastoquinone \rightarrow \mathit{cyt}_6^b \rightarrow plastocyanin \rightarrow PS\ I \rightarrow NADPH}$$

Photosystem II

PS II is an extremely complex, highly organized transmembrane structure that contains a *water-splitting complex*, chlorophylls and carotenoid pigments, a *reaction center* (P680), pheophytin (a pigment similar to chlorophyll), and two quinones. It uses the energy of sunlight to transfer electrons from water to a mobile electron carrier in the membrane called *plastoquinone*:

$$\mathbf{H_2O \rightarrow P680 \rightarrow P680^* \rightarrow plastoquinone}$$

Plastoquinone, in turn, transfers electrons to cyt_6^b, which feeds them into PS I.

The Water-Splitting Complex

The step $\mathbf{H_2^{\,o} \rightarrow P680}$ is performed by a poorly understood structure embedded within PS II called the *water-splitting complex* or the *oxygen-evolving complex*. It catalyzes a reaction that splits water into electrons, protons and oxygen:

$$\mathbf{2H_2O \rightarrow 4H^+ + 4e^- + O_2}$$

The electrons are transferred to special chlorophyll molecules (embedded in PS II) that are promoted to a higher-energy state by the energy of photons.

Photosystem II (or water-plastoquinone oxidoreductase) is the first protein complex in the light-dependent reactions of oxygenic photosynthesis. It is located in the thylakoid membrane of plants, algae, and cyanobacteria. Within the photosystem, enzymes capture photons of light to energize electrons that are then transferred through a variety of coenzymes and cofactors to reduce plastoquinone to plastoquinol. The energized electrons are replaced by oxidizing water to form hydrogen ions and molecular oxygen.

By replenishing lost electrons with electrons from the splitting of water, photosystem II provides the electrons for all of photosynthesis to occur. The hydrogen ions (protons) generated by the oxidation of water help to create a proton gradient that is used by

ATP synthase to generate ATP. The energized electrons transferred to plastoquinone are ultimately used to reduce NADP+to NADPH or are used in cyclic photophosphorylation.

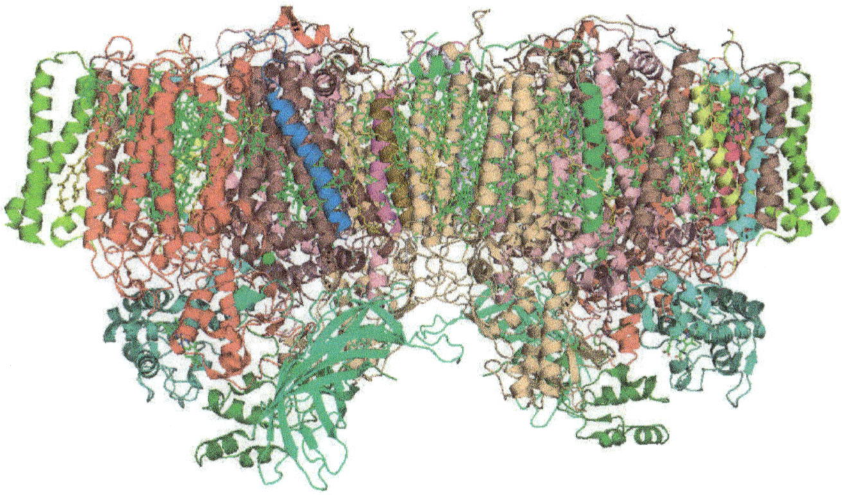

Cyanobacteria photosystem II, Dimer, PDB 2AXT

Structure

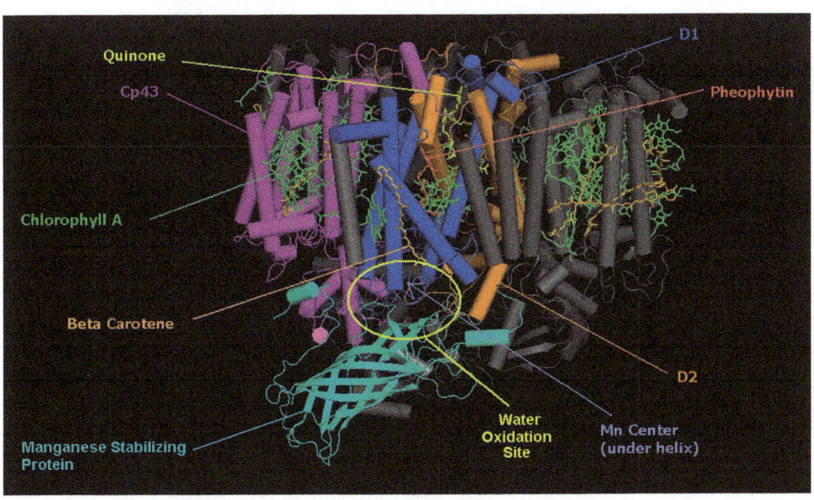

Cyanobacterial photosystem II, Monomer, PDB 2AXT.

The core of PSII consists of a pseudo-symmetric heterodimer of two homologous proteins D1 and D2. Unlike the reaction centers of all other photosystems which have a special pair of closely spaced chlorophyll molecules, the pigment that undergoes the initial photoinduced charge separation in PSII is a chlorophyll monomer. Because the positive charge is not shared across two molecules, the ionised pigment is highly oxidizing and can take part in the splitting of water.

Photosystem II (of cyanobacteria and green plants) is composed of around 20 subunits (depending on the organism) as well as other accessory, light-harvesting proteins. Each photosystem II contains at least 99 cofactors: 35 chlorophyll a, 12 beta-carotene, two pheophytin, two plastoquinone, two heme, one bicarbonate, 20 lipid, the Mn_4CaO_5 cluster (including two chloride ions), and one non heme $Fe2^+$ and two putative $Ca2^+$ ion per monomer. There are different crystal structures of photosystem II. The PDB accession codes for this protein are 3WU2, 3BZ1, 3BZ2 (3BZ1 and 3BZ2 are monomeric structures of the Photosystem II dimer), 2AXT, 1S5L, 1W5C, 1ILX, 1FE1, 1IZL.

Protein Subunits (only with known function)	
Subunit	**Function**
D1	Reaction center Protein, binds Chlorophyll P680, pheophytin, beta-carotene,quinone and manganese center
D2	Reaction center Protein
CP43	Binds manganese center
CP47	
PsbO	Manganese Stabilizing Protein
Coenzymes/Cofactors	
Molecule	**Function**
Chlorophyll	Absorbs light energy and converts it to chemical energy
Beta-carotene	quench excess photoexcitation energy
Heme b559	also Protoporphyrin IX containing iron
Pheophytin	Primary electron acceptor
Plastoquinone	Mobile intra-thylakoid membrane electron carrier
Manganese center	also known as the oxygen evolving center, or OEC

Oxygen-Evolving Complex (OEC)

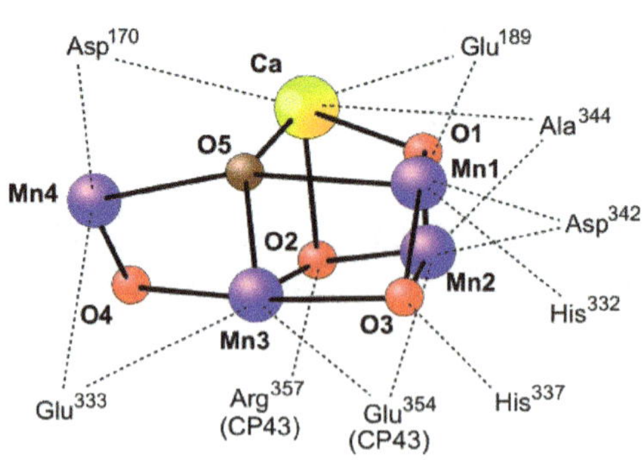

Proposed structure of Manganese Center

The oxygen-evolving complex is the site of water oxidation. It is a metallo-oxo cluster comprising four manganese ions (in oxidation states ranging from +2 to +4) and one divalent calcium ion. When it oxidizes water, producing oxygen gas and protons, it sequentially delivers the four electrons from water to a tyrosine (D1-Y161) sidechain and then to P680 itself. The structure of the oxygen-evolving complex is still contentious. The structures obtained by X-ray crystallography are particularly controversial, since there is evidence that the manganese atoms are reduced by the high-intensity X-rays used, altering the observed OEC structure. However, crystallography in combination with a variety of other (less damaging) spectroscopic methods such as EXAFS and electron paramagnetic resonance have given a fairly clear idea of the structure of the cluster. One possibility is the cubane-like structure. In 2011 the OEC of PSII was resolved to a level of 1.9 angstroms revealing five oxygen atoms serving as oxo bridges linking the five metal atoms and four water molecules bound to the Mn4CaO5 cluster; more than 1,300 water molecules were found in each photosystem II monomer, some forming extensive hydrogen-bonding networks that may serve as channels for protons, water or oxygen molecules.

Water Splitting

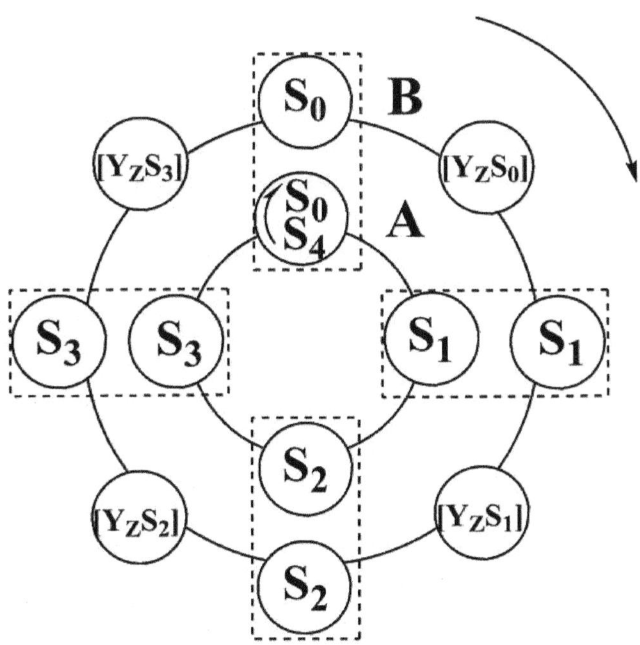

Water-splitting process: Electron transport and regulation. The first level (**A**) shows the original Kok model of the S-states cycling, the second level (**B**) shows the link between the electron transport (S-states advancement) and the relaxation process of the intermediate S-states (YzSn, n=0,1,2,3) formation

Photosynthetic water splitting (or oxygen evolution) is one of the most important reactions on the planet, since it is the source of nearly all the atmosphere's oxygen.

Moreover, artificial photosynthetic water-splitting may contribute to the effective use of sunlight as an alternative energy-source.

The mechanism of water oxidation is still not fully elucidated, but we know many details about this process. The oxidation of water to molecular oxygen requires extraction of four electrons and four protons from two molecules of water. The experimental evidence that oxygen is released through cyclic reaction of oxygen evolving complex (OEC) within one PSII was provided by Pierre Joliot et al. They have shown that, if dark-adapted photosynthetic material (higher plants, algae, and cyanobacteria) is exposed to a series of single turnover flashes, oxygen evolution is detected with typical period-four damped oscillation with maxima on the third and the seventh flash and with minima on the first and the fifth flash (for review, see). Based on this experiment, Bessel Kok and co-workers introduced a cycle of five flash-induced transitions of the so-called **S-states**, describing the four redox states of OEC: When four oxidizing equivalents have been stored (at the S_4-state), OEC returns to its basic S_0-state. In the absence of light, the OEC will "relax" to the S_1 state; the S_1 state is often described as being "dark-stable". The S_1 state is largely considered to consist of manganese ions with oxidation states of Mn^{3+}, Mn^{3+}, Mn^{4+}, Mn^{4+}. Finally, the **intermediate S-states** were proposed by Jablonsky and Lazar as a regulatory mechanism and link between S-states and tyrosine Z.

The Reaction Center

The excitation P680 → P680*of the reaction center pigment P680 occurs here. These special chlorophyll molecules embedded in PS II absorb the energy of photons, with maximal absorption at 680 nm. Electrons within these molecules are promoted to a higher-energy state. This is one of two core processes in photosynthesis, and it occurs with astonishing efficiency (greater than 90%) because, in addition to direct excitation by light at 680 nm, the energy of light first harvested by *antenna proteins* at other wavelengths in the light-harvesting system is also transferred to these special chlorophyll molecules.

This is followed by the step P680*→ pheophytin, and then on to plastoquinone, which occurs within the reaction center of PS II. High-energy electrons are transferred to plastoquinone before it subsequently picks up two protons to become plastoquinol. Plastoquinol is then released into the membrane as a mobile electron carrier.

This is the second core process in photosynthesis. The initial stages occur within *picoseconds*, with an efficiency of 100%. The seemingly impossible efficiency is due to the precise positioning of molecules within the reaction center. This is a solid-state process, not a chemical reaction. It occurs within an essentially crystalline environment created by the macromolecular structure of PS II. The usual rules of chemistry (which involve random collisions and random energy distributions) do not apply in solid-state environments.

Link of Water-Splitting Complex and Chlorophyll Excitation

When the chlorophyll passes the electron to pheophytin, it obtains an electron from P_{680}^*. In turn, P_{680}^* can oxidize the Z (or Y_Z) molecule. Once oxidized, the Z molecule can derive electrons from the oxygen-evolving complex.

Summary

PS II is a transmembrane structure found in all chloroplasts. It splits water into electrons, protons and molecular oxygen. The electrons are transferred to plastoquinone, which carries them to a proton pump. Molecular oxygen is released into the atmosphere.

The emergence of such an incredibly complex structure, a macromolecule that converts the energy of sunlight into potentially useful work with efficiencies that are impossible in ordinary experience, seems almost magical at first glance. Thus, it is of considerable interest that, in essence, the same structure is found in purple bacteria.

Cytochrome *b/6*

PS II and PS I are connected by a transmembrane proton pump, cytochrome *b/6* complex (plastoquinol—plastocyanin reductase; EC 1.10.99.1). Electrons from PS II are carried by plastoquinol to cyt*b/6*, where they are removed in a stepwise fashion (reforming plastoquinone) and transferred to a water-soluble electron carrier called *plastocyanin*. This redox process is coupled to the pumping of four protons across the membrane. The resulting proton gradient (together with the proton gradient produced by the water-splitting complex in PS II) is used to make ATP via ATP synthase.

The similarity in structure and function between cytochrome *b/6* (in chloroplasts) and cytochrome *bc/1* (*Complex III* in mitochondria) is striking. Both are transmembrane structures that remove electrons from a mobile, lipid-soluble electron carrier (plastoquinone in chloroplasts; ubiquinone in mitochondria) and transfer them to a mobile, water-soluble electron carrier (plastocyanin in chloroplasts; cytochrome *c* in mitochondria). Both are proton pumps that produce a transmembrane proton gradient.

Photosystem I

Photosystem I (PS I, or plastocyanin: ferredoxin oxidoreductase) is the second photosystem in the photosynthetic light reactions of algae, plants, and some bacteria. Photosystem I (PSI) is an integral membrane protein complex that uses light energy to produce the high energy carriers ATP and NADPH. More than 110 co-factors, significantly more than photosystem II, comprise the PS I system.

Photosystem I is named because it was discovered before photosystem II. Aspects of PS I were discovered in the 1950s, but the significances of these discoveries was not yet known. Louis Duysens first proposed the concepts of photosystems I and II in 1960, and, in the same year, a proposal by Fay Bendall and Robert Hill assembled earlier discoveries into a cohesive theory of serial photosynthetic reactions. Hill and Bendall's hypothesis was later justified in experiments conducted in 1961 by Duysens and Witt groups.

Components and Action of Photosystem I

Two main subunits of PS I, PsaA and PsaB, are closely related proteins involved in the binding of P700, A0, A1, and Fx. PsaA and PsaB are both integral membrane proteins of 730 to 750 amino acids that seem to contain 11 transmembrane segments. The Fx 4Fe-4S iron-sulphur centre is bound by four cysteines; two of these cysteines are provided by the PsaA protein and the two others by PsaB. The two cysteines in both proteins are proximal and located in a loop between the ninth and tenth transmembrane segments. A leucine zipper motif seems to be present downstream of the cysteines and could contribute to dimerisation of psaA/psaB. The terminal electron acceptors, FA and FB, are located in a 9 kDa protein called PsaC.

Protein Subunits	**proton turn into form of ATP**
Subunit	**Description**
PsaA	
PsaB	
PsaC	
PsaD	
PsaE	
PsaI	
PsaJ	
PsaK	
PsaL	
PsaM	
PsaX	
Cytochrome $\mathbf{b}_6$f complex	Soluble protein
F_a	In electron transport chain **(ETC)**
F_b	In ETC
F_x	In ETC
Ferredoxin	Electron carrier in ETC
Plastocyanin	Soluble protein
Lipids	
MGDG II	Monogalactosyldiglyceride lipid
PG I	Phosphatidylglycerol phospholipid
PG III	Phosphatidylglycerol phospholipid
PG IV	Phosphatidylglycerol phospholipid
Pigments	
Chlorophyll **a**	90 pigment **molecules in antenna system**
Chlorophyll a	5 pigment molecules in ETC
Chlorophyll a_0	Early electron acceptor of modified chlorophyll in ETC
Chlorophyll a'	1 pigment molecule in ETC
β-Carotene	22 carotenoid **pigment molecules**
Coenzymes/Cofactors	
Molecule	**Description**
Q_K-A	Early electron acceptor vitamin K_1 phylloquinone **in ETC**
Q_K-B	Early electron acceptor vitamin K_1 phylloquinone in ETC

FNR	Ferredoxin-NADP+ oxidoreductase enzyme
Ca2+	Calcium **ion**
Mg2+	Magnesium **ion**

Photon

Photoexcitation of the pigment molecules in the antenna complex induces electron transfer.

Antenna Complex

The antenna complex is composed of molecules of chlorophyll and carotenoids mounted on two proteins. These pigment molecules transmit the resonance energy from photons when they become photoexcited. Antenna molecules can absorb all wavelengths of light within the visible spectrum. The number of these pigment molecules varies from organism to organism. For instance, the cyanobacterium Synechococcus elongatus (Thermosynechococcus elongatus) has about 100 chlorophylls and 20 carotenoids, whereas spinach chloroplasts have around 200 chlorophylls and 50 carotenoids. Located within the antenna complex of PS I are molecules of chlorophyll called P700 reaction centers. The energy passed around by antenna molecules is directed to the reaction center. There may be as many as 120 or as few as 25 chlorophyll molecules per P700.

P700 Reaction Center

The P700 reaction center is composed of modified chlorophyll a that best absorbs light at a wavelength of 700nm, with higher wavelengths causing bleaching.P700 receives energy from antenna molecules and uses the energy from each photon to raise an electron to a higher energy level. These electrons are moved in pairs in an oxidation/reduction process from P700 to electron acceptors. P700 has an electric potential of about -1.2 volts. The reaction center is made of two chlorophyll molecules and is therefore referred to as a dimer. The dimer is thought to be composed of one chlorophyll a molecule and one chlorophyll a› molecule (p700, webber). However, if P700 forms a complex with other antenna molecules, it can no longer be a dimer.

Modified Chlorophyll A_0

Modified chlorophyll A_0 is an early electron acceptor in PS I. Chlorophyll A_0 accepts electrons from P700 before passing them along to another early electron acceptor.

Phylloquinone A_1

Phylloquinone A_1 is the next early electron acceptor in PS I. Phylloquinone is also sometimes called vitamin K_1. Phylloquinone A_1 oxidizes A_0 in order to receive the electron and in turn reduces F_x in order to pass the electron to F_b and F_a.

Iron-Sulfur Complex

Three proteinaceous iron-sulfur reaction centers are found in PS I. Labeled F_x, F_a, and F_b, they serve as electron relays. F_a and F_b are bound to protein subunits of the PS I complex and F_x is tied to the PS I complex. Various experiments have shown some disparity between theories of iron-sulfur co-factor orientation and operation order.

Ferredoxin

Ferredoxin (Fd) is a soluble protein that facilitates reduction of $NADP^+$ to NADPH. Fd moves to carry an electron either to a lone thylakoid or to an enzyme that reduces $NADP^+$. Thylakoid membranes have one binding site for each function of Fd. The main function of Fd is to carry an electron from the iron-sulfur complex to the enzyme ferredoxin-$NADP^+$ reductase.

Ferredoxin-NADP⁺ Reductase (FNR)

This enzyme transfers the electron from reduced ferredoxin to $NADP^+$ to complete the reduction to NADPH. FNR may also accept an electron from NADPH by binding to it.

Plastocyanin

Like the Fe-S clusters, plastocyanin is an electron relay that transfers the electron to the P700 reaction center of the PS I.

Ycf4 Protein Domain

The Ycf4 protein domain is found on the thylakoid membrane and is vital to photosystem I. This thylakoid transmembrane protein helps assemble the components of photosystem I, without it, photosynthesis would be inefficient.

Green Sulfur Bacteria and the Evolution of PS I

Molecular data show that PS I likely evolved from the photosystems of green-sulfur bacteria. The photosystems of green sulfur bacteria and those of cyanobacteria, algae, and higher plants are not the same, however there are many analogous functions and similar structures. Three main features are similar between the different photosystems. First, redox potential is negative enough to reduce ferredoxin. Next, the electron-accepting reaction centers include iron-sulfur proteins. Last, redox centres in complexes of both photosystems are constructed upon a protein subunit dimer. The photosystem of green sulfur bacteria even contains all of the same co-factors of the electron transport chain in PS I. The number and degree of similarities between the two photosystems strongly indicates that PS I is derived from the analogous photosystem of green-sulfur bacteria.

The cyclic light-dependent reactions occur when only the sole photosystem being used is photosystem I. Photosystem I excites electrons which then cycle from the transport protein, ferredoxin (Fd), to the cytochrome complex, b6f, to another transport protein, plastocyanin (Pc), and back to photosystem I. A proton gradient is created across the thylakoid membrane (6) as protons (3) are transported from the chloroplast stroma (4) to the thylakoid lumen (5). Through chemiosmosis, ATP (9) is produced where ATP synthase (1) binds an inorganic phosphate group (8) to an ADP molecule (7).

PS I accepts electrons from plastocyanin and transfers them either to NADPH (*noncyclic electron transport*) or back to cytochrome *b/6* (*cyclic electron transport*):

plastocyanin → P700 → P700* → FNR → NADPH

↑ ↓

b

6 ← plastoquinone

PS I, like PS II, is a complex, highly organized transmembrane structure that contains antenna chlorophylls, a reaction center (P700), phylloquinine, and a number of iron-sulfur proteins that serve as intermediate redox carriers.

The light-harvesting system of PS I uses multiple copies of the same transmembrane proteins used by PS II. The energy of absorbed light (in the form of delocalized, high-energy electrons) is funneled into the reaction center, where it excites special chlorophyll molecules (P700, maximum light absorption at 700 nm) to a higher energy level. The process occurs with astonishingly high efficiency.

Electrons are removed from excited chlorophyll molecules and transferred through a series of intermediate carriers to *ferredoxin*, a water-soluble electron carrier. As in PS II, this is a solid-state process that operates with 100% efficiency.

There are two different pathways of electron transport in PS I. In *noncyclic electron transport*, ferredoxin carries the electron to the enzyme ferredoxin $NADP^+$ oxidoreductase (FNR) that reduces $NADP^+$ to NADPH. In *cyclic electron transport*, electrons from ferredoxin are transferred (via plastoquinone) to a proton pump, cytochrome *b/6*. They are then returned (via plastocyanin) to P700.

NADPH and ATP are used to synthesize organic molecules from CO_2. The ratio of NADPH to ATP production can be adjusted by adjusting the balance between cyclic and noncyclic electron transport.

It is noteworthy that PS I closely resembles photosynthetic structures found in green sulfur bacteria, just as PS II resembles structures found in purple bacteria.

Photosynthetic Electron Transport Chains in Bacteria

PS II, PS I, and cytochrome*b/6* are found in chloroplasts. All plants and all photosynthetic algae contain chloroplasts, which produce NADPH and ATP by the mechanisms described above. In essence, the same transmembrane structures are also found in *cyanobacteria*.

Unlike plants and algae, cyanobacteria are prokaryotes. They do not contain chloroplasts. Rather, they bear a striking resemblance to chloroplasts themselves. This suggests that organisms resembling cyanobacteria were the evolutionary precursors of chloroplasts. One imagines primitive eukaryotic cells taking up cyanobacteria as intracellular symbionts in a process known as endosymbiosis.

Cyanobacteria

Cyanobacteria contain structures similar to PS II and PS I in chloroplasts. Their light-harvesting system is different from that found in plants (they use *phycobilins*, rather than chlorophylls, as antenna pigments), but their electron transport chain

$$\begin{array}{lcccccccccccccc} H_2O & \rightarrow & \text{PS II} & \rightarrow & \text{plastoquinone} & \rightarrow & {}^{b}_{6} & \rightarrow & \text{cytochrome } {}^{c}_{6} & \rightarrow & \text{PS I} & \rightarrow & \text{ferredoxin} & \rightarrow & \text{NADPH} \\ & & & & & & & & \uparrow & & & & \downarrow & & \\ & & & & & & & & {}^{b}_{6} & \leftarrow & & & \text{plastoquinone} & & \end{array}$$

is, in essence, the same as the electron transport chain in chloroplasts. The mobile water-soluble electron carrier is cytochrome *c/6* in cyanobacteria, plastocyanin in plants.

Cyanobacteria can also synthesize ATP by oxidative phosphorylation, in the manner of other bacteria. The electron transport chain is

$$\text{NADH dehydrogenase} \rightarrow \text{plastoquinone} \rightarrow {}^{b}_{6} \rightarrow \text{cytochrome } {}^{c}_{6} \rightarrow \text{cytochrome } {}^{aa}_{3} \rightarrow O_2$$

where the mobile electron carriers are plastoquinone and cytochrome *c/6*, while the proton pumps are NADH dehydrogenase, *b/6* and cytochrome *aa/3*.

Cyanobacteria are the only bacteria that produce oxygen during photosynthesis. Earth's primordial atmosphere was anoxic. Organisms like cyanobacteria produced our present-day oxygen-containing atmosphere.

The other two major groups of photosynthetic bacteria, purple bacteria and green sulfur bacteria, contain only a single photosystem and do not produce oxygen.

Purple Bacteria

Purple bacteria contain a single photosystem that is structurally related to PS II in cyanobacteria and chloroplasts:

$$\text{P870} \rightarrow \text{P870}^{*} \rightarrow \text{ubiquinone} \rightarrow bc/1 \rightarrow \text{cytochrome } c/2 \quad P870$$

This is a *cyclic* process in which electrons are removed from an excited chlorophyll molecule (*bacteriochlorophyll*; P870), passed through an electron transport chain to a proton pump (cytochrome *bc/1* complex, similar but not identical to cytochrome *bc/1* in chloroplasts), and then returned to the chlorophyll molecule. The result is a proton gradient, which is used to make ATP via ATP synthase. As in cyanobacteria and chloroplasts, this is a solid-state process that depends on the precise orientation of various functional groups within a complex transmembrane macromolecular structure.

To make NADPH, purple bacteria use an external electron donor (hydrogen, hydrogen sulfide, sulfur, sulfite, or organic molecules such as succinate and lactate) to feed electrons into a reverse electron transport chain.

Green Sulfur Bacteria

Green sulfur bacteria contain a photosystem that is analogous to PS I in chloroplasts:

```
  P840 → P840* → ferredoxin → NADH
   ↑                   ↓
cyt c553 ← bc1 ← menaquinone
```

There are two pathways of electron transfer. In *cyclic electron transfer*, electrons are removed from an excited chlorophyll molecule, passed through an electron transport chain to a proton pump, and then returned to the chlorophyll. The mobile electron carriers are, as usual, a lipid-soluble quinone and a water-soluble cytochrome. The resulting proton gradient is used to make ATP.

In *noncyclic electron transfer*, electrons are removed from an excited chlorophyll molecule and used to reduce NAD^+ to NADH. The electrons removed from P840 must be replaced. This is accomplished by removing electrons from H_2S, which is oxidized to sulfur (hence the name "green *sulfur* bacteria").

Purple bacteria and green sulfur bacteria occupy relatively minor ecological niches in the present day biosphere. They are of interest because of their importance in precambrian ecologies, and because their methods of photosynthesis were the likely evolutionary precursors of those in modern plants.

The first ideas about light being used in photosynthesis were proposed by Colin Flannery in 1779 who recognized it was sunlight falling on plants that was required, although Joseph Priestley had noted the production of oxygen without the association with light in 1772. Cornelius Van Niel proposed in 1931 that photosynthesis is a case of general mechanism where a photon of light is used to photo decompose a hydrogen donor and the hydrogen being used to reduce CO_2. Then in 1939, Robin Hill showed that isolated chloroplasts would make oxygen, but not fix CO_2 showing the light and dark reactions occurred in different places. Although they are referred to as light and dark reactions, both of them take place only in the presence of light. This led later to the discovery of photosystems I and II.

LIGHT-INDEPENDENT REACTIONS

The light-independent reactions of photosynthesis are chemical reactions that convert carbon dioxide and other compounds into glucose. These reactions occur in the stroma, the fluid-filled area of a chloroplast outside of the thylakoid membranes. These reactions take the products (ATP and NADPH) of light-dependent reactions and perform further chemical processes on them. There are three phases to the light-independent reactions, collectively called the Calvin cycle: carbon fixation, reduction reactions, and ribulose 1,5-bisphosphate (RuBP) regeneration.

Despite its name, this process occurs only when light is available. Plants do not carry out the Calvin cycle during nighttime. They instead release sucrose into the phloem from their starch reserves. This process happens when light is available independent of the kind of photosynthesis (C3 carbon fixation, C4 carbon fixation, and Crassulacean acid metabolism); CAM plants store malic acid in their vacuoles every night and release it by day in order to make this process work.They are also known as dark reactions.

Coupling to other Metabolic Pathways

These reactions are closely coupled to the thylakoid electron transport chain as reducing power provided by NADPH produced in the photosystem I is actively needed. The process of photorespiration, also known as C2 cycle, is also coupled to the dark reactions, as it results from an alternative reaction of the Rubisco enzyme, and its final byproduct is also another glyceraldehyde-3-P.

The Calvin cycle, Calvin–Benson–Bassham (CBB) cycle, reductive pentose phosphate cycle or C3 cycle is a series of biochemical redox reactions that take place in the stroma of chloroplast in photosynthetic organisms. It is also known as the light-independent reactions.

The cycle was discovered by Melvin Calvin, James Bassham, and Andrew Benson at the University of California, Berkeley by using the radioactive isotope carbon-14. It is one of the light-independent reactions used for carbon fixation.

Photosynthesis occurs in two stages in a cell. In the first stage, light-dependent reactions capture the energy of light and use it to make the energy-storage and transport molecules ATP and NADPH. The light-independent Calvin cycle uses the energy from short-lived electronically excited carriers to convert carbon dioxide and water into organic compounds that can be used by the organism (and by animals that feed on it). This set of reactions is also called *carbon fixation*. The key enzyme of the cycle is called RuBisCO. In the following biochemical equations, the chemical species (phosphates and carboxylic acids) exist in equilibria among their various ionized states as governed by the pH.

The enzymes in the Calvin cycle are functionally equivalent to most enzymes used in other metabolic pathways such as gluconeogenesis and the pentose phosphate pathway, but they are to be found in the chloroplast stroma instead of the cell cytosol, separating the reactions. They are activated in the light (which is why the name «dark reaction» is misleading), and also by products of the light-dependent reaction. These regulatory functions prevent the Calvin cycle from being respired to carbon dioxide. Energy (in the form of ATP) would be wasted in carrying out these reactions that have no net productivity.

The sum of reactions in the Calvin cycle is the following:

$3\ CO_2 + 6\ NADPH + 5\ H_2O + 9\ ATP \rightarrow$ glyceraldehyde-3-phosphate (G3P) $+ 2\ H^+ + 6\ NADP^+ + 9\ ADP + 8\ P_i$ (P_i = inorganic phosphate)

Hexose (six-carbon) sugars are not a product of the Calvin cycle. Although many texts list a product of photosynthesis as $C_6H_{12}O_6$, this is mainly a convenience to counter the equation of respiration, where six-carbon sugars are oxidized in mitochondria. The carbohydrate products of the Calvin cycle are three-carbon sugar phosphate molecules, or "triose phosphates," namely, glyceraldehyde-3-phosphate (G3P).

Steps

In the first stage of the Calvin cycle, a CO_2 molecule is incorporated into one of two three-carbon molecules (glyceraldehyde 3-phosphate or G3P), where it uses up two molecules of ATP and two molecules of NADPH, which had been produced in the light-dependent stage. The three steps involved are:

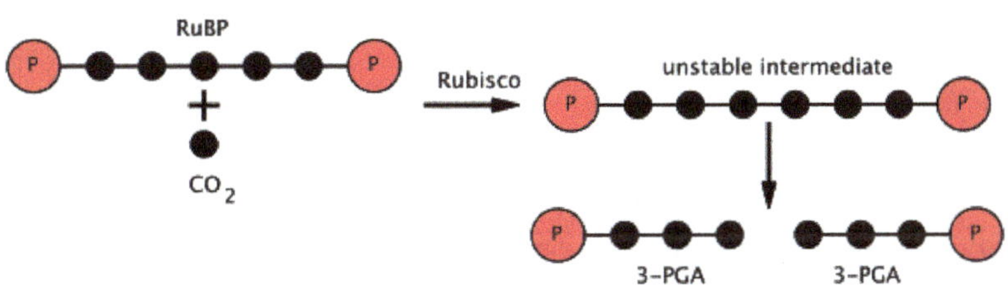

Calvin cycle step 1 (black circles represent carbon atoms)

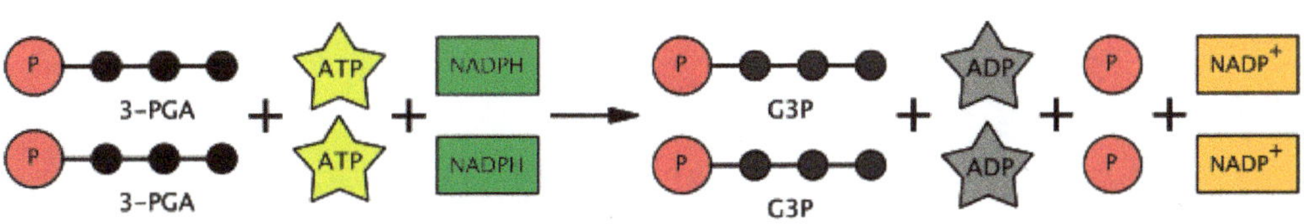

Calvin cycle steps 2 and 3 combined

- The enzyme RuBisCO catalyses the carboxylation of ribulose-1,5-bisphosphate, RuBP, a 5-carbon compound, by carbon dioxide (a total of 6 carbons) in a two-step reaction. The product of the first step is enediol-enzyme complex that can capture CO_2 or O_2. Thus, enediol-enzyme complex is the real carboxylase/ oxygenase. The CO_2 that is captured by enediol in second step produces a six-carbon intermediate initially that immediately splits in half, forming two molecules of 3-phosphoglycerate, or 3-PGA, a 3-carbon compound (also: 3-phosphoglyceric acid, PGA, 3PGA).
- The enzyme phosphoglycerate kinase catalyses the phosphorylation of 3-PGA by ATP (which was produced in the light-dependent stage). 1,3-Bisphosphoglycerate

(1,3BPGA, glycerate-1,3-bisphosphate) and ADP are the products. (However, note that two 3-PGAs are produced for every CO_2 that enters the cycle, so this step utilizes two ATP per CO_2 fixed.)

- The enzyme glyceraldehyde 3-phosphate dehydrogenase catalyses the reduction of 1,3BPGA by NADPH (which is another product of the light-dependent stage). Glyceraldehyde 3-phosphate (also called G3P, GP, TP, PGAL, GAP) is produced, and the NADPH itself is oxidized and becomes $NADP^+$. Again, two NADPH are utilized per CO_2 fixed.

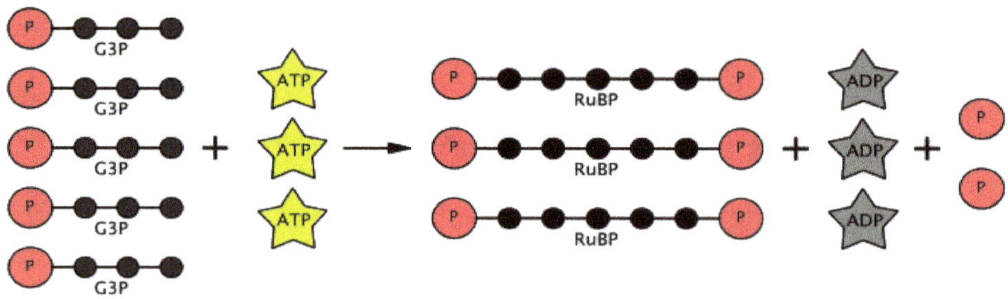

Regeneration stage of the Calvin cycle

The next stage in the Calvin cycle is to regenerate RuBP. Five G3P molecules produce three RuBP molecules, using up three molecules of ATP. Since each CO_2 molecule produces two G3P molecules, three CO_2molecules produce six G3P molecules, of which five are used to regenerate RuBP, leaving a net gain of one G3P molecule per three CO_2molecules (as would be expected from the number of carbon atoms involved).

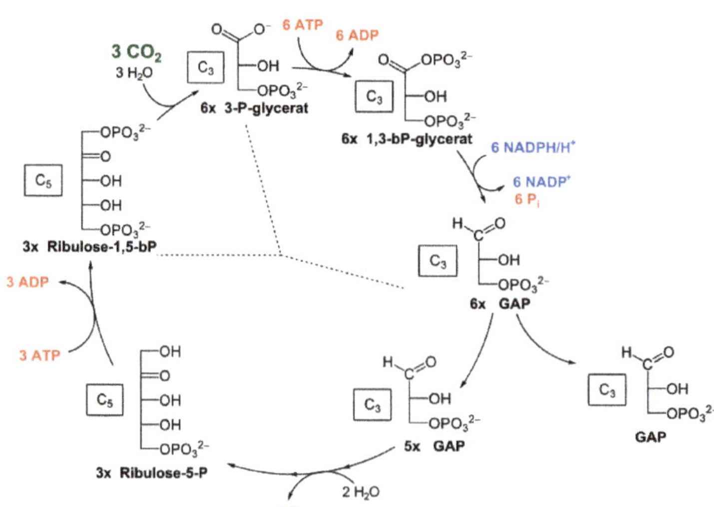

Simplified C3 cycle with structural formulas

The regeneration stage can be broken down into steps.

- Triose phosphate isomerase converts all of the G3P reversibly into dihydroxyacetone phosphate (DHAP), also a 3-carbon molecule.
- Aldolase and fructose-1,6-bisphosphatase convert a G3P and a DHAP into fructose 6-phosphate (6C). A phosphate ion is lost into solution.
- Then fixation of another CO_2 generates two more G3P.
- F6P has two carbons removed by transketolase, giving erythrose-4-phosphate. The two carbons on transketolase are added to a G3P, giving the ketose xylulose-5-phosphate (Xu5P).
- E4P and a DHAP (formed from one of the G3P from the second CO_2 fixation) are converted into sedoheptulose-1,7-bisphosphate (7C) by aldolase enzyme.
- Sedoheptulose-1,7-bisphosphatase (one of only three enzymes of the Calvin cycle that are unique to plants) cleaves sedoheptulose-1,7-bisphosphate into sedoheptulose-7-phosphate, releasing an inorganic phosphate ion into solution.
- Fixation of a third CO_2 generates two more G3P. The ketose S7P has two carbons removed by transketolase, giving ribose-5-phosphate (R5P), and the two carbons remaining on transketolase are transferred to one of the G3P, giving another Xu5P. This leaves one G3P as the product of fixation of 3 CO_2, with generation of three pentoses that can be converted to Ru5P.
- R5P is converted into ribulose-5-phosphate (Ru5P, RuP) by phosphopentose isomerase. Xu5P is converted into RuP by phosphopentose epimerase.
- Finally, phosphoribulokinase (another plant-unique enzyme of the pathway) phosphorylates RuP into RuBP, ribulose-1,5-bisphosphate, completing the Calvin *cycle*. This requires the input of one ATP.

Thus, of six G3P produced, five are used to make three RuBP (5C) molecules (totaling 15 carbons), with only one G3P available for subsequent conversion to hexose. This requires nine ATP molecules and six NADPH molecules per three CO_2 molecules. The equation of the overall Calvin cycle is shown diagrammatically below.

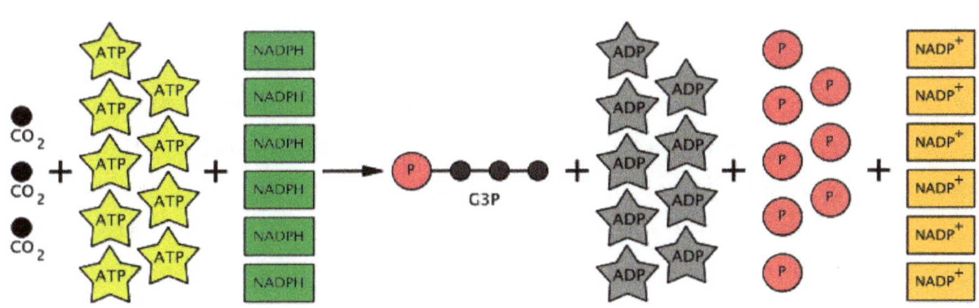

The overall equation of the Calvin cycle (black circles represent carbon atoms)

RuBisCO also reacts competitively with O_2 instead of CO_2 in photorespiration. The rate of photorespiration is higher at high temperatures. Photorespiration turns RuBP into 3-PGA and 2-phosphoglycolate, a 2-carbon molecule that can be converted via glycolate and glyoxalate to glycine. Via the glycine cleavage system and tetrahydrofolate, two glycines are converted into serine +CO_2. Serine can be converted back to 3-phosphoglycerate. Thus, only 3 of 4 carbons from two phosphoglycolates can be converted back to 3-PGA. It can be seen that photorespiration has very negative consequences for the plant, because, rather than fixing CO_2, this process leads to loss of CO_2. C4 carbon fixation evolved to circumvent photorespiration, but can occur only in certain plants native to very warm or tropical climates—corn, for example.

Products

The immediate products of one turn of the Calvin cycle are 2 glyceraldehyde-3-phosphate (G3P) molecules, 3 ADP, and 2 $NADP^+$. (ADP and $NADP^+$ are not really "products." They are regenerated and later used again in the Light-dependent reactions). Each G3P molecule is composed of 3 carbons. In order for the Calvin cycle to continue, RuBP (ribulose 1,5-bisphosphate) must be regenerated. So, 5 out of 6 carbons from the 2 G3P molecules are used for this purpose. Therefore, there is only 1 net carbon produced to play with for each turn. To create 1 surplus G3P requires 3 carbons, and therefore 3 turns of the Calvin cycle. To make one glucose molecule (which can be created from 2 G3P molecules) would require 6 turns of the Calvin cycle. Surplus G3P can also be used to form other carbohydrates such as starch, sucrose, and cellulose, depending on what the plant needs.

Light-Dependent Regulation

Despite its widespread names (both light-independent and dark reactions), these reactions do not occur in the dark or at night. There is a light-dependent regulation of the cycle enzymes, as the third step requires reduced NADP; and this process would be a waste of energy, as there is no electron flow in the dark.

There are 2 regulation systems at work when the cycle needs to be turned on or off: thioredoxin/ferredoxin activation system, which activates some of the cycle enzymes, and the RuBisCo enzyme activation, active in the Calvin cycle, which involves its own activase.

The thioredoxin/ferredoxin system activates the enzymes glyceraldehyde-3-P dehydrogenase, glyceraldehyde-3-P phosphatase, fructose-1,6-bisphosphatase, sedoheptulose-1,7-bisphosphatase, and ribulose-5-phosphatase kinase, which are key points of the process. This happens when light is available, as the ferredoxin protein is reduced in the photosystem I complex of the thylakoid electron chain when electrons are circulating through it. Ferredoxin then binds to and reduces the thioredoxin protein, which activates the cycle enzymes by severing a cystine bond found in all these enzymes. This is a dynamic process as the same bond is formed again by other proteins that deactivate the enzymes. The implications of this process are that the enzymes remain mostly activated by day and are deactivated in the dark when there is no more reduced ferredoxin available.

The enzyme RuBisCo has its own activation process, which involves a more complex process. It is necessary that a specific lysine amino acid be carbamylated in order to activate the enzyme. This lysine binds to RuBP and leads to a non-functional state if left uncarbamylated. A specific activase enzyme, called RuBisCo activase, helps this carbamylation process by removing one proton from the lysine and making the binding of the carbon dioxide molecule possible. Even then the RuBisCo enzyme is not yet functional, as it needs a magnesium ion to be bound to the lysine in order to function. This magnesium ion is released from the thylakoid lumen when the inner pH drops due to the active pumping of protons from the electron flow. RuBisCo activase itself is activated by increased concentrations of ATP in the stroma caused by its phosphorylation.

REFERENCES

1. Vass, Imre; Cser, Krisztian (April 2009). "Janus-faced charge recombinations in photosystem II photoinhibition". Trends in Plant Science. 14 (4): 200–205. doi:10.1016/j.tplants.2009.01.009. ISSN 1360-1385. PMID 19303349. Retrieved 2012-10-30.
2. Keren, Nir; Berg, Alexander; van Kan, Paul J. M.; Levanon, Haim; Ohad, Itzhak (1997-02-18). "Mechanism of photosystem II photoinactivation and D1 protein degradation at low light: The role of back electron flow". Proceedings of the National Academy of Sciences. 94 (4): 1579–1584. doi:10.1073/pnas.94.4.1579. ISSN 0027-8424. PMC 19834 . PMID 11038602. Retrieved 2012-10-30.
3. Rajni Govindjee. "The Z-Scheme Diagram of Photosynthesis". Retrieved March 2, 2006.
4. "Photosynthesis". McGraw Hill Encyclopedia of Science and Technology. 2007. p. 472.
5. Ingenhousz, J (1779). Experiments Upon Vegetables. London: Elmsly and Payne.
6. Priestley, J (1772). "Observations on Different Kinds of Air". 62. London: Phil. Trans. Roy. Soc.: 147–264. doi:10.1098/rstl.1772.0021.
7. van Niel, C. B. (1931). "On the morphology and physiology of the purple and green sulfur bacteria". Arch. Microbiol. 3: 1–114. doi:10.1007/BF00454965.
8. Hill, R. (May 1939). "Oxygen Produced by Isolated Chloroplasts". Proceedings of the Royal Society of London. Series B, Biological Sciences. 127 (847): 192–210. doi:10.1098/rspb.1939.0017.

CHAPTER 5

Photodissociation in Photosynthesis

Chapter Contents

- **Introduction**
- **Photolysis in Photosynthesis**
- **Photocatalysis**
- **Photochemistry**

Photons are tiny bits of light and other kinds of electromagnetic radiation. Photons can sometimes break apart molecules. When this happens, it is called photodissociation. When a photon runs into a molecule, it adds energy to the molecule. Molecules have chemical bonds that hold the atoms in them together with each other. If the chemical bond breaks, the molecule falls apart. Sometimes photons have enough energy to break the bonds in a molecule. A photon of ultraviolet (UV) "light" has more energy than a photon of visible light. UV photons can cause photodissociation more easily than visible light photons can.

INTRODUCTION

Photodissociation, photolysis, or photodecomposition is a chemical reaction in which a chemical compound is broken down by photons. It is defined as the interaction of one or more photons with one target molecule. Photodissociation is not limited to visible light. Any photon with sufficient energy can affect the chemical bonds of a chemical compound. Since a photon's energy is inversely proportional to its wavelength, electromagnetic waves with the energy of visible light or higher, such as ultraviolet light, x-rays and gamma rays are usually involved in such reactions.

PHOTOLYSIS IN PHOTOSYNTHESIS

Photolysis is part of the light-dependent reactions of photosynthesis. The general reaction of photosynthetic photolysis can be given as

$H_2A + 2$ photons (light) $\rightarrow 2\ e^- + 2\ H^+ + A$

The chemical nature of «A» depends on the type of organism. In purple sulfur bacteria, hydrogen sulfide (H_2S) is oxidized to sulfur (S). In oxygenic photosynthesis, water (H_2O) serves as a substrate for photolysis resulting in the generation of diatomic oxygen (O_2). This is the process which returns oxygen to Earth's atmosphere. Photolysis of water occurs in the thylakoids of cyanobacteria and the chloroplasts of green algae and plants.

Energy Transfer Models

The conventional, semi-classical, model describes the photosynthetic energy transfer process as one in which excitation energy hops from light-capturing pigment molecules to reaction center molecules step-by-step down the molecular energy ladder.

The effectiveness of photons of different wavelengths depends on the absorption spectra of the photosynthetic pigments in the organism. Chlorophylls absorb light in the violet-blue and red parts of the spectrum, while accessory pigments capture other wavelengths as well. The phycobilins of red algae absorb blue-green light which penetrates deeper into water than red light, enabling them to photosynthesize in deep waters. Each absorbed photon causes the formation of an exciton (an electron excited to a higher energy state) in the pigment molecule. The energy of the exciton is transferred to a chlorophyll molecule (P680, where P stands for pigment and 680 for its absorption maximum at 680 nm) in the reaction center of photosystem II via resonance energy transfer. P680 can also directly absorb a photon at a suitable wavelength.

Photolysis during photosynthesis occurs in a series of light-driven oxidation events. The energized electron (exciton) of P680 is captured by a primary electron acceptor of the photosynthetic electron transfer chain and thus exits photosystem II. In order to repeat the reaction, the electron in the reaction center needs to be replenished. This occurs by oxidation of water in the case of oxygenic photosynthesis. The electron-deficient reaction center of photosystem II (P680*) is the strongest biological oxidizing agent yet discovered, which allows it to break apart molecules as stable as water.

The water-splitting reaction is catalyzed by the oxygen evolving complex of photosystem II. This protein-bound inorganic complex contains four manganese ions, plus calcium and chloride ions as cofactors. Two water molecules are complexed by the manganese cluster, which then undergoes a series of four electron removals (oxidations) to replenish the reaction center of photosystem II. At the end of this cycle, free oxygen (O_2) is generated and the hydrogen of the water molecules has been converted to four protons released into the thylakoid lumen.

These protons, as well as additional protons pumped across the thylakoid membrane coupled with the electron transfer chain, form a proton gradient across the membrane that drives photophosphorylation and thus the generation of chemical energy in the form of adenosine triphosphate (ATP). The electrons reach the P700 reaction center of photosystem

I where they are energized again by light. They are passed down another electron transfer chain and finally combine with the coenzyme $NADP^+$ and protons outside the thylakoids to NADPH. Thus, the net oxidation reaction of water photolysis can be written as:

$2\ H_2O + 2\ NADP^+ + 8$ photons (light) $\rightarrow 2\ NADPH + 2\ H^+ + O_2$

The free energy change (ΔG) for this reaction is 102 kilocalories per mole. Since the energy of light at 700 nm is about 40 kilocalories per mole of photons, approximately 320 kilocalories of light energy are available for the reaction. Therefore, approximately one-third of the available light energy is captured as NADPH during photolysis and electron transfer. An equal amount of ATP is generated by the resulting proton gradient. Oxygen as a byproduct is of no further use to the reaction and thus released into the atmosphere.

Quantum Models

In 2007 a quantum model was proposed by Graham Fleming and his co-workers which includes the possibility that photosynthetic energy transfer might involve quantum oscillations, explaining its unusually high efficiency.

According to Fleming there is direct evidence that remarkably long-lived wavelike electronic quantum coherence plays an important part in energy transfer processes during photosynthesis, which can explain the extreme efficiency of the energy transfer because it enables the system to sample all the potential energy pathways, with low loss, and choose the most efficient one. This claim has, however, since been proven wrong in several publications .

This approach has been further investigated by Gregory Scholes and his team at the University of Toronto, which in early 2010 published research results that indicate that some marine algae make use of quantum-coherent electronic energy transfer (EET) to enhance the efficiency of their energy harnessing.

Photolysis in the Atmosphere

Photolysis occurs in the atmosphere as part of a series of reactions by which primary pollutants such as hydrocarbons and nitrogen oxides react to form secondary pollutants such as peroxyacyl nitrates. See photochemical smog.

The two most important photodissociaton reactions in the troposphere are firstly:

$O_3 + h\nu \rightarrow O_2 + O(^1D)\ \lambda < 320$ nm

which generates an excited oxygen atom which can react with water to give the hydroxyl radical:

$O(^1D) + H_2O \rightarrow 2\ {}^{\bullet}OH$

The hydroxyl radical is central to atmospheric chemistry as it initiates the oxidation of hydrocarbons in the atmosphere and so acts as a detergent.

Secondly the reaction:

$NO_2 + h\nu \rightarrow NO + O$

is a key reaction in the formation of tropospheric ozone.

The formation of the ozone layer is also caused by photodissociation. Ozone in the Earth›s stratosphere is created by ultraviolet light striking oxygen molecules containing two oxygen atoms (O_2), splitting them into individual oxygen atoms (atomic oxygen). The atomic oxygen then combines with unbroken O_2 to create ozone, O_3. In addition, photolysis is the process by which CFCs are broken down in the upper atmosphere to form ozone-destroying chlorine free radicals.

Astrophysics

In astrophysics, photodissociation is one of the major processes through which molecules are broken down (but new molecules are being formed). Because of the vacuum of the interstellar medium, molecules and free radicals can exist for a long time. Photodissociation is the main path by which molecules are broken down. Photodissociation rates are important in the study of the composition of interstellar clouds in which stars are formed.

Examples of photodissociation in the interstellar medium are (hv is the energy of a single photon of frequency v)

$$H_2O + h\nu \rightarrow H + OH$$

$$CH_4 + h\nu \rightarrow CH_3 + H$$

Atmospheric Gamma-Ray Bursts

Currently orbiting satellites detect an average of about one gamma-ray burst per day. Because gamma-ray bursts are visible to distances encompassing most of the observable universe, a volume encompassing many billions of galaxies, this suggests that gamma-ray bursts must be exceedingly rare events per galaxy.

Measuring the exact rate of gamma-ray bursts is difficult, but for a galaxy of approximately the same size as the Milky Way, the expected rate (for long GRBs) is about one burst every 100,000 to 1,000,000 years. Only a few percent of these would be beamed towards Earth. Estimates of rates of short GRBs are even more uncertain because of the unknown beaming fraction, but are probably comparable.

A gamma-ray burst in the Milky Way, if close enough to Earth and beamed towards it, could have significant effects on the biosphere. The absorption of radiation in the atmosphere would cause photodissociation of nitrogen, generating nitric oxide that would act as a catalyst to destroy ozone.

The atmospheric photodissociation

N_2 -> 2N

O_2 -> 2O

CO_2 -> C + 2O

H_2O -> 2H + O

$2NH_3$ -> $3H_2$ + N_2

would yield

- NO_2 (consumes up to 400 ozone molecules)
- CH_2 (nominal)
- CH_4 (nominal)
- CO_2

(incomplete)

According to a 2004 study, a GRB at a distance of about a kiloparsec could destroy up to half of Earth's ozone layer; the direct UV irradiation from the burst combined with additional solar UV radiation passing through the diminished ozone layer could then have potentially significant impacts on the food chain and potentially trigger a mass extinction. The authors estimate that one such burst is expected per billion years, and hypothesize that the Ordovician-Silurian extinction event could have been the result of such a burst.

There are strong indications that long gamma-ray bursts preferentially or exclusively occur in regions of low metallicity. Because the Milky Way has been metal-rich since before the Earth formed, this effect may diminish or even eliminate the possibility that a long gamma-ray burst has occurred within the Milky Way within the past billion years. No such metallicity biases are known for short gamma-ray bursts. Thus, depending on their local rate and beaming properties, the possibility for a nearby event to have had a large impact on Earth at some point in geological time may still be significant.

Multiple Photon Dissociation

Single photons in the infrared spectral range usually are not energetic enough for direct photodissociation of molecules. However, after absorption of multiple infrared photons a molecule may gain internal energy to overcome its barrier for dissociation. Multiple photon dissociation (MPD, IRMPD with infrared radiation) can be achieved by applying high power lasers, e.g. a carbon dioxide laser, or a free electron laser, or by long interaction times of the molecule with the radiation field without the possibility for rapid cooling, e.g. by collisions. The latter method allows even for MPD induced by black body radiation, a technique called blackbody infrared radiative dissociation (BIRD).

Flash Photolysis

Flash photolysis is a pump-probe laboratory technique, in which a sample is firstly excited by a strong pulse (called pump pulse) of light from a laser of nanosecond, picosecond, or femtosecond pulse width or by a short-pulse light source such as a flash lamp. This first strong pulse starts a chemical reaction or leads to an increased population for energy levels other than the ground state within a sample of atoms or molecules. Typically the absorption of light by the sample is recorded within short time intervals (by a so-called test pulses) to monitor relaxation or reaction processes initiated by the pump pulse.

Flash photolysis was developed shortly after World War II as a result of the military's attempts to build cameras fast enough to photograph missiles in flight. The technique was developed in 1949 by Manfred Eigen, Ronald George Wreyford Norrish and George Porter, who won the 1967 Nobel Prize in Chemistry for this invention. Over the next 40 years the technique became more powerful and sophisticated due to developments in optics and lasers. Also, the interest in this method grew considerably as the practical applications expanded from chemistry to areas such as biology, materials science, and environmental sciences. Today flash photolysis facilities are extensively used by researchers to study light-induced processes in organic molecules, polymers, nanoparticles, semiconductors, photosynthesis in plants, signaling, and light-induced conformational changes in biological systems.

PHOTOCATALYSIS

In chemistry, photocatalysis is the acceleration of a photoreaction in the presence of a catalyst. In catalysed photolysis, light is absorbed by an adsorbed substrate. In photogenerated catalysis, the photocatalytic activity (PCA) depends on the ability of the catalyst to create electron–hole pairs, which generate free radicals (e.g. hydroxyl radicals: •OH) able to undergo secondary reactions. Its practical application was made possible by the discovery of water electrolysis by means of titanium dioxide.

In the experiment above, photons from a light source (out of frame on the right hand side) are absorbed by the surface of the titanium dioxide disc, exciting electrons within the material. These then react with the water molecules, splitting it into its constituents of hydrogen and oxygen. In this experiment, chemicals dissolved in the water prevent the formation of oxygen, which would otherwise recombine with the hydrogen.

Types of Photocatalysis

Homogeneous Photocatalysis

In homogeneous photocatalysis, the reactants and the photocatalysts exist in the same phase. The most commonly used homogeneous photocatalysts include ozone and photo-Fenton systems (Fe^+ and Fe^+/H_2O_2). The reactive species is the •OH which is used for different purposes. The mechanism of hydroxyl radical production by ozone can follow two paths.

$O_3 + h\nu \rightarrow O_2 + O(1D)$ (? O_3 "-" $h\nu \rightarrow O_2 + O(1D)$?)

$O(1D) + H_2O \rightarrow \bullet OH + \bullet OH$

$O(1D) + H_2O \rightarrow H_2O_2$

$H_2O_2 + h\nu \rightarrow \bullet OH + \bullet OH$

Similarly, the Fenton system produces hydroxyl radicals by the following mechanism

$Fe^{2+} + H_2O_2 \rightarrow HO\bullet + Fe^{3+} + OH^-$

$Fe^{3+} + H_2O_2 \rightarrow Fe^{2+} + HO\bullet 2 + H^+$

$Fe^{2+} + HO\bullet \rightarrow Fe^{3+} + OH^-$

In photo-Fenton type processes, additional sources of OH radicals should be considered: through photolysis of H_2O_2, and through reduction of Fe^{3+} ions under UV light:

$H_2O_2 + h\nu \rightarrow HO\bullet + HO\bullet$

$Fe^{3+} + H_2O + h\nu \rightarrow Fe^{2+} + HO\bullet + H^+$

The efficiency of Fenton type processes is influenced by several operating parameters like concentration of hydrogen peroxide, pH and intensity of UV. The main advantage of this process is the ability of using sunlight with light sensitivity up to 450 nm, thus avoiding the high costs of UV lamps and electrical energy. These reactions have been proven more efficient than the other photocatalysis but the disadvantages of the process are the low pH values which are required, since iron precipitates at higher pH values and the fact that iron has to be removed after treatment.

Heterogeneous Photocatalysis

Heterogeneous catalysis has the catalyst in a different phase from the reactants. Heterogeneous photocatalysis is a discipline which includes a large variety of reactions: mild or total oxidations, dehydrogenation, hydrogen transfer, $^{18}O_2$–$^{16}O_2$ and deuterium-alkane isotopic exchange, metal deposition, water detoxification, gaseous pollutant removal, etc.

Most common heterogeneous photocatalysts are transition metal oxides and semiconductors, which have unique characteristics. Unlike the metals which have a continuum of electronic states, semiconductors possess a void energy region where no energy levels are available to promote recombination of an electron and hole produced by photoactivation in the solid. The void region, which extends from the top of the filled valence band to the bottom of the vacant conduction band, is called the band gap. When a photon with energy equal to or greater than the materials band gap is absorbed by the semiconductor, an electron is excited from the valence band to the conduction band, generating a positive hole in the valence band. The excited electron and hole can recombine and release the energy gained from the excitation of the electron as heat. Recombination is undesirable and leads to an inefficient photocatalyst. The ultimate goal of the process is to have a reaction between the excited electrons with an oxidant to produce a reduced product, and also a reaction between the generated holes with a reductant to produce an oxidized product. Due to the generation of positive holes and electrons, oxidation-reduction reactions take place at the surface of semiconductors. In the oxidative reaction, the positive holes react with the moisture present on the surface and produce a hydroxyl radical.

Oxidative reactions due to photocatalytic effect:

$UV + MO \rightarrow MO\ (h + e^-)$

Here MO stands for metal oxide ---

$h^+ + H_2O \rightarrow H^+ + \bullet OH$

$2\ h^+ + 2\ H_2O \rightarrow 2\ H^+ + H_2O_2$

$H_2O_2 \rightarrow 2\ \bullet OH$

The reductive reaction due to photocatalytic effect:

$e^- + O_2 \rightarrow \bullet O_2^-$

$\bullet O_2 - + HO\bullet 2 + H^+ \rightarrow H_2O_2 + O_2$

$HOOH \rightarrow HO\bullet + \bullet OH$

Ultimately, the hydroxyl radicals are generated in both the reactions. These hydroxyl radicals are very oxidative in nature and non selective with redox potential of (E_0 = +3.06 V)

Applications

- Conversion of water to hydrogen gas by photocatalytic water splitting. An efficient photocatalyst in the UV range is based on a sodium tantalite ($NaTaO_3$) doped with La and loaded with a cocatalyst nickel oxide. The surface of the sodium tantalite crystals is grooved with so called nanosteps that is a result of doping with lanthanum (3–15 nm range, see nanotechnology). The NiO particles which facilitate hydrogen gas evolution are present on the edges, with the oxygen gas evolving from the grooves.
- Use of titanium dioxide in self-cleaning glass. Free radicals generated from TiO_2 oxidize organic matter.
- Disinfection of water by supported titanium dioxide photocatalysts, a form of solar water disinfection (SODIS).
- Use of titanium dioxide in self-sterilizing photocatalytic coatings (for application to food contact surfaces and in other environments where microbial pathogens spread by indirect contact).
- Oxidation of organic contaminants using magnetic particles that are coated with titanium dioxide nanoparticles and agitated using a magnetic field while being exposed to UV light.
- Conversion of carbon dioxide into gaseous hydrocarbons using titanium dioxide in the presence of water. As an efficient absorber in the UV range, titanium dioxide nanoparticles in the anatase and rutile phases are able to generate excitons by promoting electrons across the band gap. The electrons and holes react with the surrounding water vapor to produce hydroxyl radicals and protons. At present,

proposed reaction mechanisms usually suggest the creation of a highly reactive carbon radical from carbon monoxide and carbon dioxide which then reacts with the photogenerated protons to ultimately form methane. Although the efficiencies of present titanium dioxide based photocatalysts are low, the incorporation of carbon based nanostructures such as carbon nanotubes and metallic nanoparticles have been shown to enhance the efficiency of these photocatalysts.

- Sterilization of surgical instruments and removal of unwanted fingerprints from sensitive electrical and optical components.
- A less-toxic alternative to tin and copper-based antifouling marine paints, ePaint, generates hydrogen peroxide by photocatalysis.
- Decomposition of crude oil with TiO_2 nanoparticles: by using titanium dioxide photocatalysts and UV-A radiation from the sun, the hydrocarbons found in crude oil can be turned into H_2O and CO_2. Higher amounts of oxygen and UV radiation increased the degradation of the model organics. These particles can be placed on floating substrates, making it easier to recover and catalyze the reaction. This is relevant since oil slicks float on top of the ocean and photons from the sun target the surface more than the inner depth of the ocean. By covering floating substrates like woodchips with epoxy adhesives, water logging can be prevented and TiO_2 particles can stick to the substrates. With more research, this method should be applicable to other organics.
- Decontamination of water with photocatalysis and adsorption: the removal and destruction of organic contaminants in groundwater can be addressed through the impregnation of adsorbents with photoactive catalysts. These adsorbents attract contaminating organic atoms/molecules like tetrachloroethylene to them. The photoactive catalysts impregnated inside speed up the degradation of the organics. Adsorbents are placed in packed beds for 18 hours, which would attract and degrade the organic compounds. The spent adsorbents would then be placed in regeneration fluid, essentially taking away all organics still attached by passing hot water counter-current to the flow of water during the adsorption process to speed up the reaction. The regeneration fluid then gets passed through the fixed beds of silica gel photocatalysts to remove and decompose the rest of the organics left. Through the use of fixed bed reactors, the regeneration of adsorbents can help increase the efficiency.
- Decomposition of polyaromatic hydrocarbons (PAHs). Triethylamine (TEA) was utilized to solvate and extract the polyaromatic hydrocarbons (PAHs) found in crude oil. By solvating these PAHs, TEA can attract the PAHs to itself. Once removed, TiO_2 slurries and UV light can photocatalytically degrade the PAHs. The figure shows the high success rate of this experiment. With high yielding of recoveries of 93–99% of these contaminants, this process has become an innovative idea that can be finalized for actual environmental usage. This procedure demonstrates the ability to develop photocatalysts that would be performed at ambient pressure, ambient temperature, and at a cheaper cost.

Quantification of Photocatalytic Activity

ISO 22197 -1:2007 specifies a test method for the determination of the nitric oxide removal performance of materials that contain a photocatalyst or have photocatalytic films on the surface.

Specific FTIR systems are used to characterise photocatalytic activity and/or passivity especially with respect to Volatile Organic Compounds VOCs and representative matrices of the binders applied.

Recent studies show that mass spectrometry can be a powerful tool to determine photocatalytic activity of certain materials by following the decomposition of gaseous pollutants such as nitrogen oxides or carbon dioxide

Photohydrogen

Photohydrogen is hydrogen produced with the help of artificial or natural light This is how the leaf of a tree splits water molecules into protons (hydrogen ions), electrons (to make carbohydrates) and oxygen (released into the air as a waste product). Photohydrogen may also be produced by the photodissociation of water by ultraviolet light.

Photohydrogen is sometimes discussed in the context of obtaining renewable energy from sunlight, by using microscopic organisms such as bacteria or algae. These organisms create hydrogen with the help of hydrogenase enzymes which convert protons derived from the water splitting reaction into hydrogen gas which can then be collected and used as a biofuel.

PHOTOCHEMISTRY

Photochemistry is the branch of chemistry concerned with the chemical effects of light. Generally, this term is used to describe a chemical reaction caused by absorption of ultraviolet (wavelength from 100 to 400 nm), visible light (400 – 750 nm) or infrared radiation (750 – 2500 nm).

In nature, photochemistry is of immense importance as it is the basis of photosynthesis, vision, and the formation of vitamin D with sunlight. Photochemical reactions proceed differently than temperature-driven reactions. Photochemical paths access high energy intermediates that cannot be generated thermally, thereby overcoming large activation barriers in a short period of time, and allowing reactions otherwise inaccessible by thermal processes. Photochemistry is also destructive, as illustrated by the photodegradation of plastics.

Photochemical immersion well reactor (50 mL) with a mercury-vapor lamp.

Basics of Photochemistry

Grotthuss–Draper Law and Stark-Einstein Law

Photoexcitation is the first step in a photochemical process where the reactant is elevated to a state of higher energy, an excited state. The first law of photochemistry, known as the Grotthuss–Draper law (for chemists Theodor Grotthuss and John W. Draper), states that light must be absorbed by a chemical substance in order for a photochemical reaction to take place. According to the second law of photochemistry, known as the Stark-Einstein law (for physicists Johannes Stark and Albert Einstein), for each photon of light absorbed by a chemical system, no more than one molecule is activated for a photochemical reaction, as defined by the quantum yield.

Fluorescence and Phosphorescence

When a molecule in the ground state (S_0) absorbs light, one electron is excited to a higher orbital level. This electron maintains its spin according to the spin selection rule; other transitions would violate the law of conservation of angular momentum. The excitation to a higher singlet state can be from HOMO to LUMO or to a higher orbital, so that singlet excitation states S_1, S_2, S_3... at different energies are possible.

Kasha's rule stipulates that higher singlet states would quickly relax by radiationless decay or internal conversion (IC) to S_1. Thus, S_1 is usually, but not always, the only relevant singlet excited state. This excited state S_1 can further relax to S_0 by IC, but also by an allowed radiative transition from S_1 to S_0 that emits a photon; this process is called fluorescence.

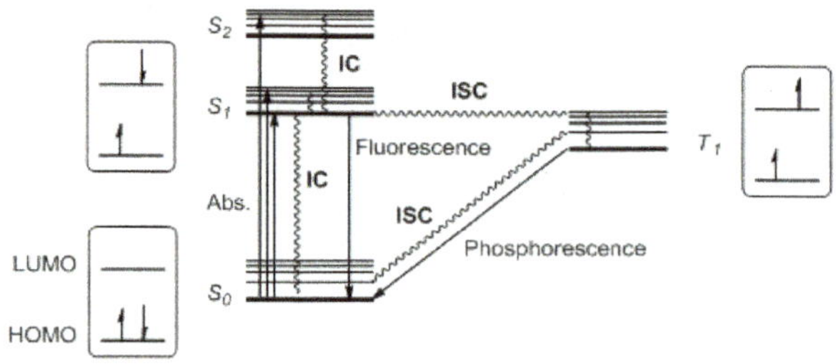

Jablonski diagram. Radiative paths are represented by straight arrows and non-radiative paths by curly lines.

Alternatively, it is possible for the excited state S_1 to undergo spin inversion and to generate a triplet excited state T_1 having two unpaired electrons with the same spin. This violation of the spin selection rule is possible by intersystem crossing (ISC) of the vibrational and electronic levels of S_1 and T_1. According to Hund's rule of maximum multiplicity, this T_1 state would be somewhat more stable than S_1.

This triplet state can relax to the ground state S_0 by radiationless IC or by a radiation pathway called phosphorescence. This process implies a change of electronic spin, which is forbidden by spin selection rules, making phosphorescence (from T_1 to S_0) much slower than fluorescence (from S_1 to S_0). Thus, triplet states generally have longer lifetimes than singlet states. These transitions are usually summarized in a state energy diagram or Jablonski diagram, the paradigm of molecular photochemistry.

These excited species, either S_1 or T_1, have a half empty low-energy orbital, and are consequently more oxidizing than the ground state. But at the same time, they have an electron in a high energy orbital, and are thus more reducing. In general, excited species are prone to participate in electron transfer processes.

Experimental Set-Up

Photochemical immersion well reactor (750 mL) with a mercury-vapor lamp

Photochemical reactions require a light source that emits wavelengths corresponding to an electronic transition in the reactant. In the early experiments (and in everyday life), sunlight was the light source, although it is polychromatic. Mercury-vapor lamps are more common in the laboratory. Low pressure mercury vapor lamps mainly emit at 254 nm. For polychromatic sources, wavelength ranges can be selected using filters. Alternatively, laser beams are usually monochromatic (although two or more wavelengths can be obtained using nonlinear optics) and LEDs have a relatively narrowband that can be efficiently used, as well as Rayonet lamps, to get approximately monochromatic beams.

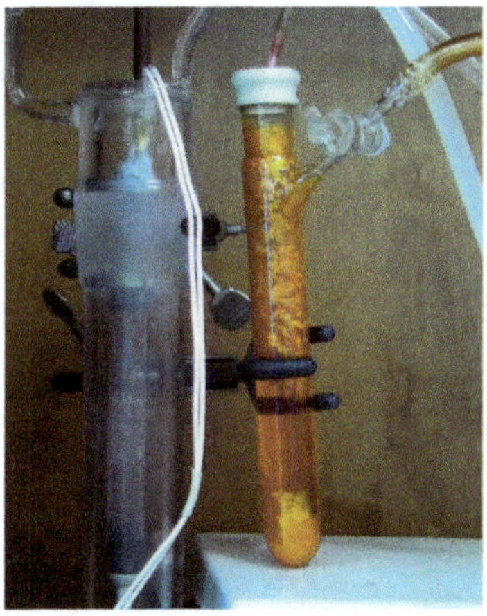

Schlenk tube containing slurry of orange crystals of $Fe_2(CO)_9$ in acetic acid after its photochemical synthesis from $Fe(CO)_5$. The mercury lamp (connected to white power cords) can be seen on the left, set inside a water-jacketed quartz tube.

The emitted light must of course reach the targeted functional group without being blocked by the reactor, medium, or other functional groups present. For many applications, quartz is used for the reactors as well as to contain the lamp. Pyrex absorbs at wavelengths shorter than 275 nm. The solvent is an important experimental parameter. Solvents are potential reactants and for this reason, chlorinated solvents are avoided because the C-Cl bond can lead to chlorination of the substrate. Strongly absorbing solvents prevent photons from reaching the substrate. Hydrocarbon solvents absorb only at short wavelengths and are thus preferred for photochemical experiments requiring high energy photons. Solvents containing unsaturation absorb at longer wavelengths and can usefully filter out short wavelengths. For example, cyclohexane and acetone "cut off" (absorb strongly) at wavelengths shorter than 215 and 330 nm, respectively.

Photochemistry in Combination with Flow Chemistry

Continuous flow photochemistry offers multiple advantages over batch photochemistry. Photochemical reactions are driven by the number of photons that are able to activate molecules causing the desired reaction. The large surface area to volume ratio of a microreactor maximizes the illumination, and at the same time allows for efficient cooling, which decreases the thermal side products.

Principles

In the case of photochemical reactions, light provides the activation energy. Simplistically, light is one mechanism for providing the activation energy required for many reactions. If laser light is employed, it is possible to selectively excite a molecule so as to produce a desired electronic and vibrational state. Equally, the emission from a particular state may be selectively monitored, providing a measure of the population of that state. If the chemical system is at low pressure, this enables scientists to observe the energy distribution of the products of a chemical reaction before the differences in energy have been smeared out and averaged by repeated collisions.

The absorption of a photon of light by a reactant molecule may also permit a reaction to occur not just by bringing the molecule to the necessary activation energy, but also by changing the symmetry of the molecule's electronic configuration, enabling an otherwise inaccessible reaction path, as described by the Woodward–Hoffmann selection rules. A 2+2 cycloaddition reaction is one example of a pericyclic reaction that can be analyzed using these rules or by the related frontier molecular orbital theory.

Some photochemical reactions are several orders of magnitude faster than thermal reactions; reactions as fast as 10^{-9} seconds and associated processes as fast as 10^{-15} seconds are often observed.

The photon can be absorbed directly by the reactant or by a photosensitizer, which absorbs the photon and transfers the energy to the reactant. The opposite process is called quenching when a photoexited state is deactivated by a chemical reagent.

Most photochemical transformations occur through a series of simple steps known as primary photochemical processes. One common example of these processes is the excited state proton transfer.

Photochemical Reactions

Examples of Photochemical Reactions

- **Photosynthesis**: plants use solar energy to convert carbon dioxide and water into glucose and oxygen.
- Human formation of vitamin D by exposure to sunlight.
- **Bioluminescence**: *e.g.* In fireflies, an enzyme in the abdomen catalyzes a reaction that produced light.
- Polymerizations started by photoinitiators, which decompose upon absorbing light to produce the free radicals for radical polymerization.
- Photodegradation of many substances, e.g. polyvinyl chloride and Fp. Medicine bottles are often made with darkened glass to prevent the drugs from photodegradation.
- **Photodynamic therapy**: light is used to destroy tumors by the action of singlet oxygen generated by photosensitized reactions of triplet oxygen. Typical photosensitizers include tetraphenylporphyrin and methylene blue. The resulting

singlet oxygen is an aggressive oxidant, capable of converting C-H bonds into C-OH groups.

- Photoresist technology, used in the production of microelectronic components.
- Vision is initiated by a photochemical reaction of rhodopsin.
- Toray photochemical production of ε-caprolactame.
- Photochemical production of artemisinin, anti-malaria drug.

Organic Photochemistry

Examples of photochemical organic reactions are electrocyclic reactions, radical reactions, photoisomerization and Norrish reactions.

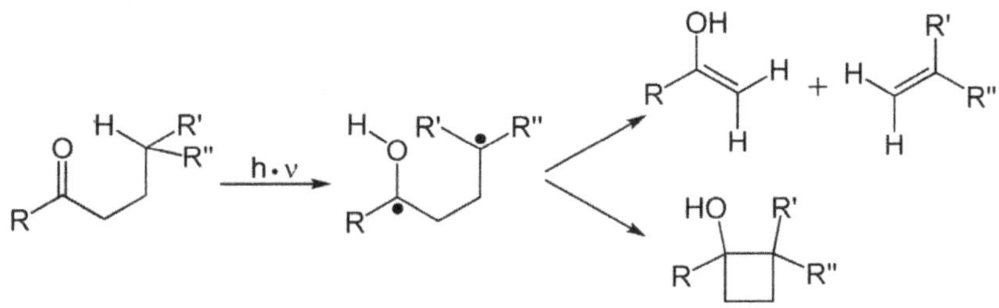

Alkenes undergo many important reactions that proceed via a photon-induced π to π^* transition. The first electronic excited state of an alkene lack the π-bond, so that rotation about the C-C bond is rapid and the molecule engages in reactions not observed thermally. These reactions include cis-trans isomerization, cycloaddition to other (ground state) alkene to give cyclobutane derivatives. The cis-trans isomerization of a (poly) alkene is involved in retinal, a component of the machinery of vision. The dimerization of alkenes is relevant to the photodamage of DNA, where thymine dimers are observed upon illuminating DNA to UV radiation. Such dimers interfere with transcription. The beneficial effects of sunlight are associated with the photochemically induced retro-cyclization (decyclization) reaction of ergosterol to give vitamin D. In the DeMayo reaction, an alkene reacts with a 1,3-diketone reacts via its enol to yield a 1,5-diketone. Still another common photochemical reaction is Zimmerman's Di-pi-methane rearrangement.

In an industrial application, about 100,000 tonnes of benzyl chloride are prepared annually by the gas-phase photochemical reaction of toluene with chlorine. The light is absorbed by chlorine molecule, the low energy of this transition being indicated by the yellowish color of the gas. The photon induces homolysis of the Cl-Cl bond, and the resulting chlorine radical converts toluene to the benzyl radical:

$Cl_2 + h\nu \rightarrow 2\ Cl\cdot$

$C_6H_5CH_3 + Cl\cdot \rightarrow C_6H_5CH_2\cdot + HCl$

$C_6H_5CH_2\cdot + Cl\cdot \rightarrow C_6H_5CH_2Cl$

Mercaptans can be produced by photochemical addition of hydrogen sulfide (H_2S) to alpha olefins.

Inorganic and Organometallic Photochemistry

Coordination complexes and organometallic compounds are also photoreactive. These reactions can entail cis-trans isomerization. More commonly photoreactions result in dissociation of ligands, since the photon excites an electron on the metal to an orbital that is antibonding with respect to the ligands. Thus, metal carbonyls that resist thermal substitution undergo decarbonylation upon irradiation with UV light. UV-irradiation of a THF solution of molybdenum hexacarbonyl gives the THF complex, which is synthetically useful:

$$Mo(CO)_6 + THF \rightarrow Mo(CO)_5(THF) + CO$$

In a related reaction, photolysis of iron pentacarbonyl affords diiron nonacarbonyl (see figure):

$$2\ Fe(CO)_5 \rightarrow Fe_2(CO)_9 + CO$$

Although bleaching has long been practiced, the first photochemical reaction was described by Trommsdorf in 1834. He observed that crystals of the compound α-santonin when exposed to sunlight turned yellow and burst. In a 2007 study the reaction was described as a succession of three steps taking place within a single crystal.

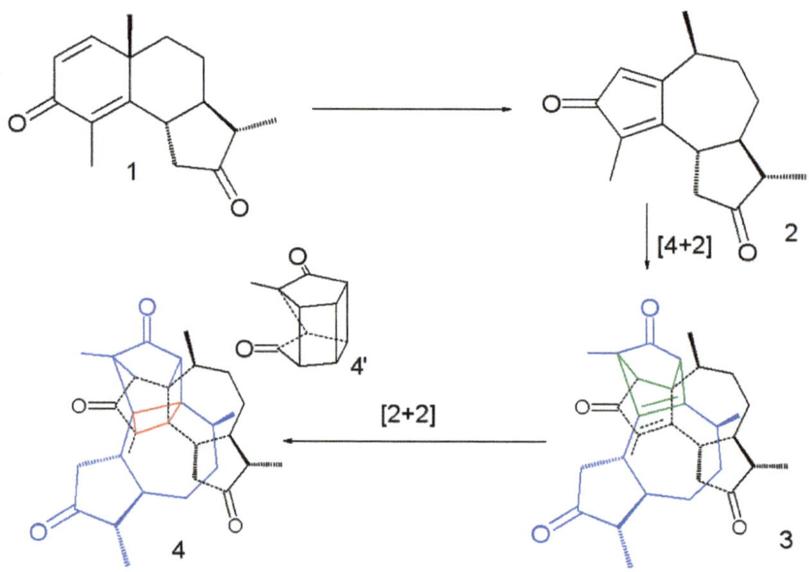

The first step is a rearrangement reaction to a cyclopentadienone intermediate **2**, the second one a dimerization in a Diels-Alder reaction (**3**) and the third one an intramolecular 2+2 cycloaddition (**4**). The bursting effect is attributed to a large change in crystal volume on dimerization.

REFERENCES

1. Campbell, Neil A.; Reece, Jane B. (2005). Biology (7th ed.). San Francisco: Pearson - Benjamin Cummings. pp. 186–191. ISBN 0-8053-7171-0.

2. Raven, Peter H.; Ray F. Evert; Susan E. Eichhorn (2005). Biology of Plants (7th ed.). New York: W.H. Freeman and Company Publishers. pp. 115–127. ISBN 0-7167-1007-2.
3. Gregory S. Engel, Tessa R. Calhoun, Elizabeth L. Read, Tae-Kyu Ahn, Tomáš Mančal, Yuan-Chung Cheng, Robert E. Blankenship and Graham R. Fleming, "Evidence for wavelike energy transfer through quantum coherence in photosynthetic systems", in Nature 446, 782-786 (12 April 2007)
4. http://www.physorg.com/news95605211.html Quantum secrets of photosynthesis revealed
5. R. Tempelaar; T. L. C. Jansen; J. Knoester (2014). "Vibrational Beatings Conceal Evidence of Electronic Coherence in the FMO Light-Harvesting Complex". J. Phys. Chem. B. 118: 12865–12872. doi:10.1021/jp510074q.
6. N. Christenson; H. F. Kauffmann; T. Pullerits; T. Mancal (2012). "Origin of Long-Lived Coherences in Light-Harvesting Complexes". J. Phys. Chem. B. 116: 7449–7454.
7. E. Thyrhaug; K. Zidek; J. Dostal; D. Bina; D. Zigmantas (2016). "Exciton Structure and Energy Transfer in the Fenna−Matthews− Olson Complex". J. Phys. Chem. Lett. 7: 1653–1660. doi:10.1021/acs.jpclett.6b00534.
8. A. G. Dijkstra; Y. Tanimura (2012). "The role of the environment time scale in light-harvesting efficiency and coherent oscillations". New J. Phys. 14: 073027. doi:10.1088/1367-2630/14/7/073027.
9. D. M. Monahan; L. Whaley-Mayda; A. Ishizaki; G. R. Fleming (2015). "Influence of weak vibrational-electronic couplings on 2D electronic spectra and inter-site coherence in weakly coupled photosynthetic complexes". J. Chem. Phys. 143: 065101. doi:10.1063/1.4928068.
10. Scholes Group Research
11. Gregory D. Scholes (7 January 2010), "Quantum-coherent electronic energy transfer: Did Nature think of it first?", Journal of Physical Chemistry Letters, 1 (1): 2–8, doi:10.1021/jz900062f
12. Elisabetta Collini; Cathy Y. Wong; Krystyna E. Wilk; Paul M. G. Curmi; Paul Brumer; Gregory D. Scholes (4 February 2010), "Coherently wired light-harvesting in photosynthetic marine algae at ambient temperature", Nature, 463 (7281), Bibcode:2010Natur.463..644C, doi:10.1038/nature08811, PMID 20130647
13. Podsiadlowski 2004
14. Guetta 2006
15. Thorsett 1995
16. Melott 2004
17. Wanjek 2005
18. Stanek 2006
19. Ejzak 2007

Carbon Fixation: The C2, C3, C4 Pathways

Chapter Contents

- **Introduction**
- **Net Vs Gross CO_2 Fixation**
- **Overview of Pathways**
- **Photorespiration**

Carbon Fixation is a process that involves fixing a carbon into place. Just like you take that piece of paper and tack it to the bulletin board so that it won't float around or end up on the floor, plants want to take carbon dioxide (CO_2) out of the air and tack it to carbohydrates so it's not floating around all over the place. The process of taking inorganic carbon (usually CO_2) and tacking it to an organic molecule (usually a carbohydrate) is known as carbon fixation.

INTRODUCTION

Carbon fixation or carbon assimilation is the conversion process of inorganic carbon (carbon dioxide) to organic compounds by living organisms. The most prominent example is photosynthesis, although chemosynthesis is another form of carbon fixation that can take place in the absence of sunlight. Organisms that grow by fixing carbon are called autotrophs. Autotrophs include photoautotrophs, which synthesize organic compounds using the energy of sunlight, and lithoautotrophs, which synthesize organic compounds using the energy of inorganic oxidation. Heterotrophs are organisms that grow using the carbon fixed by autotrophs. The organic compounds are used by heterotrophs to produce energy and to build body structures. «Fixed carbon», «reduced carbon», and «organic carbon» are equivalent terms for various organic compounds.

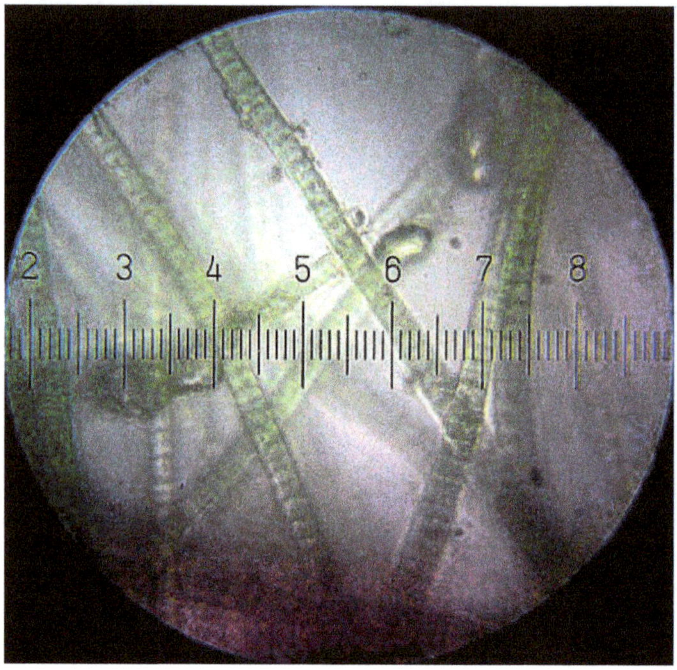

Cyanobacteria such as these carry out photosynthesis. Their emergence foreshadowed the evolution of many photosynthetic plants, which oxygenated Earth's atmosphere.

NET VS GROSS CO_2 FIXATION

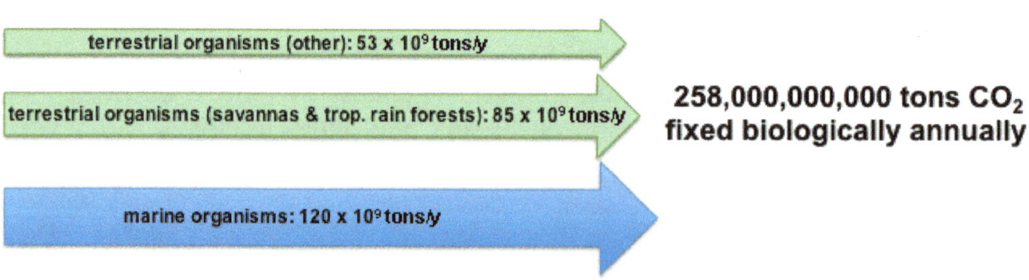

Graphic showing net annual amounts of CO_2 fixation by land and sea-based organisms.

It is estimated that approximately 258 billion tons of carbon dioxide are converted by photosynthesis annually. The majority of the fixation occurs in marine environments, especially areas of high nutrients. The gross amount of carbon dioxide fixed is much larger since approximately 40% is consumed by respiration following photosynthesis.

Given the scale of this process, it is understandable that RuBisCO is the most abundant protein on Earth.

OVERVIEW OF PATHWAYS

Six autotrophic carbon fixation pathways are known as of 2011. The Calvin cycle fixes carbon in the chloroplasts of plants and algae, and in the cyanobacteria. It also fixes carbon in the anoxygenic photosynthetic proteobacteria called purple bacteria, and in some non-phototrophic proteobacteria.

Oxygenic Photosynthesis

In photosynthesis, energy from sunlight drives the carbon fixation pathway. *Oxygenic* photosynthesis is used by the primary producers—plants, algae, and cyanobacteria. They contain the pigment chlorophyll, and use the Calvin cycle to fix carbon autotrophically. The process works like this:

$$2H_2O \rightarrow 4e^- + 4H^+ + O_2$$

$$CO_2 + 4e^- + 4H^+ \rightarrow CH_2O + H_2O$$

In the first step, water is dissociated into electrons, protons, and free oxygen. This allows the use of water, one of the most abundant substances on Earth, as an electron donor—as a source of reducing power. The release of free oxygen is a side-effect of enormous consequence. The first step uses the energy of sunlight to oxidize water to O_2, and, ultimately, to produce ATP

$$ADP + P_i \rightleftharpoons ATP + H_2O$$

and the reductant, NADPH

$$NADP^+ + 2e^- + 2H^+ \rightleftharpoons NADPH + H^+$$

In the second step, called the Calvin cycle, the actual fixation of carbon dioxide is carried out. This process consumes ATP and NADPH. The Calvin cycle in plants accounts for the preponderance of carbon fixation on land. In algae and cyanobacteria, it accounts for the preponderance of carbon fixation in the oceans. The Calvin cycle converts carbon dioxide into sugar, as triose phosphate (TP), which is glyceraldehyde 3-phosphate (GAP) together with dihydroxyacetone phosphate (DHAP):

$$3\ CO_2 + 12\ e^- + 12\ H^+ + P_i \rightarrow TP + 4\ H_2O$$

An alternative perspective accounts for NADPH (source of e^-) and ATP:

$$3\ CO_2 + 6\ NADPH + 6\ H^+ + 9\ ATP + 5\ H_2O \rightarrow TP + 6\ NADP^+ + 9\ ADP + 8\ P_i$$

The formula for inorganic phosphate (P_i) is $HOPO_3^{2-} + 2H^+$. Formulas for triose and TP are $C_2H_3O_2\text{-}CH_2OH$ and $C_2H_3O_2\text{-}CH_2OPO_3^{2-} + 2H^+$

Evolutionary Considerations

Somewhere between 3.5 and 2.3 billion years ago, the ancestors of cyanobacteria evolved oxygenic photosynthesis, enabling the use of the abundant yet relatively oxidized molecule

H_2O as an electron donor to the electron tranport chain of light-catalyzed proton-pumping responsible for efficient ATP synthesis. When this evolutionary breakthrough occurred, autotrophy (growth using inorganic carbon as the sole carbon source) is believed to have already been developed. However, the proliferation of cyanobacteria, due to their novel ability to exploit water as a source of electrons, radically altered the global environment by oxygenating the atmosphere and by achieving large fluxes of CO_2 consumption.

Carbon Concentrating Mechanisms

Many photosynthetic organisms have acquired inorganic carbon concentrating mechanisms (CCM), which increase the concentration of carbon dioxide available to the initial carboxylase of the Calvin cycle, the enzyme RuBisCO. The benefits of CCM include increased tolerance to low external concentrations of inorganic carbon, and reduced losses to photorespiration. CCM can make plants more tolerant of heat and water stress.

Carbon concentrating mechanisms use the enzyme carbonic anhydrase (CA), which catalyze both the dehydration of bicarbonate to carbon dioxide and the hydration of carbon dioxide to bicarbonate

$$HCO_3^- + H^+ \rightleftharpoons CO_2 + H_2O$$

Lipid membranes are much less permeable to bicarbonate than to carbon dioxide. To capture inorganic carbon more effectively, some plants have adapted the anaplerotic reactions

$$HCO_3^- + H^+ + PEP \rightarrow OAA + P_i$$

catalyzed by PEP carboxylase (PEPC), to carboxylate phosphoenolpyruvate (PEP) to oxaloacetate (OAA) which is a C_4 dicarboxylic acid.

CAM Plants

CAM plants that use Crassulacean acid metabolism as an adaptation for arid conditions. CO_2 enters through the stomata during the night and is converted into the 4-carbon compound, malic acid, which releases CO_2 for use in the Calvin cycle during the day, when the stomata are closed. The jade plant (*Crassula ovata*) and cacti are typical of CAM plants. Sixteen thousand species of plants use CAM. These plants have a carbon isotope signature of −20 to −10 ‰.

C_4 Plants

C_4 plants preface the Calvin cycle with reactions that incorporate CO_2 into one of the 4-carbon compounds, malic acid or aspartic acid. C_4 plants have a distinctive internal leaf anatomy. Tropical grasses, such as sugar cane and maize are C_4 plants, but there are many broadleaf plants that are C_4. Overall, 7600 species of terrestrial plants use C_4 carbon fixation, representing around 3% of all species. These plants have a carbon isotope signature of −16 to −10 ‰.

C_3 Plants

The large majority of plants are C_3 plants. They are so-called to distinguish them from the CAM and C_4 plants, and because the carboxylation products of the Calvin cycle are 3-carbon compounds. They lack C_4 dicarboxylic acid cycles, and therefore have higher carbon dioxide compensation points than CAM or C_4 plants. C_3plants have a carbon isotope signature of −24 to −33‰.

Other Autotrophic Pathways

Of the five other autotrophic pathways, two are known only in bacteria, two only in archaea, and one in both bacteria and archaea.

Reductive Citric Acid Cycle

The reductive citric acid cycle is the oxidative citric acid cycle run in reverse. It has been found in anaerobic and microaerobic bacteria. It was proposed in 1966 by Evans, Buchanan and Arnon who were working with the anoxygenic photosynthetic green sulfur bacterium that they called *Chlorobium thiosulfatophilum*. The reductive citric acid cycle is sometimes called the Arnon-Buchanan cycle.

Reductive Acetyl CoA Pathway

The reductive acetyl CoA pathway operated in strictly anaerobic bacteria (acetogens) and archaea (methanogens). The pathway was proposed in 1965 by Ljungdahl and Wood. They were working with the gram-positive acetic acid producing bacterium *Clostridium thermoaceticum*, which is now named *Moorella thermoacetica*. Hydrogenotrophic methanogenesis, which is only found in certain archaea and accounts for 80% of global methanogenesis, is also based on the reductive acetyl CoA pathway. The pathway is often referred to as the Wood–Ljungdahl pathway.

3-Hydroxypropionate and two Related Cycles

The 3-hydroxypropionate cycle is utilized *only* by green nonsulfur bacteria. It was proposed in 2002 for the anoxygenic photosynthetic *Chloroflexus aurantiacus*. None of the enzymes that participate in the 3-hydroxypropionate cycle are especially oxygen sensitive.

A variant of the 3-hydroxypropionate pathway was found to operate in aerobic extreme thermoacidophile archaeon *Metallosphaera sedula*. This pathway, called the 3-hydroxypropionate/4-hydroxybutyrate cycle . And yet another variant of the 3-hydroxypropionate pathway is the dicarboxylate/4-hydroxybutyrate cycle. It was discovered in anaerobic archaea. It was proposed in 2008 for the hyperthermophile archeon *Ignicoccus hospitalis*.

Chemosynthesis

Chemosynthesis is carbon fixation driven by the oxidation of inorganic substances (e.g., hydrogen gas or hydrogen sulfide). Sulfur- and hydrogen-oxidizing bacteria often use the Calvin cycle or the reductive citric acid cycle.

Non-Autotrophic Pathways

Although almost all heterotrophs cannot synthesize complete organic molecules from carbon dioxide, some carbon dioxide is incorporated in their metabolism.Notably pyruvate carboxylase consumes carbon dioxide (as bicarbonate ions) as part of gluconeogenesis, and carbon dioxide is consumed in various anaplerotic reactions.

Carbon Isotope Discrimination

Some carboxylases, particularly RuBisCO, preferentially bind the lighter carbon stable isotope carbon-12 over the heavier carbon-13. This is known as carbon isotope discrimination and results in carbon-12 to carbon-13 ratios in the plant that are higher than in the free air. Measurement of this ratio is important in the evaluation of water use efficiency in plants, and also in assessing the possible or likely sources of carbon in global carbon cycle studies.

PHOTORESPIRATION

Photorespiration (also known as the oxidative photosynthetic carbon cycle, or C2 photosynthesis) refers to a process in plant metabolism where the enzyme RuBisCO oxygenates RuBP, causing some of the energy produced by photosynthesis to be wasted. The desired reaction is the addition of carbon dioxide to RuBP (carboxylation), a key step in the Calvin–Benson cycle, however approximately 25% of reactions by RuBisCO instead add oxygen to RuBP (oxygenation), creating a product that cannot be used within the Calvin–Benson cycle. This process reduces the efficiency of photosynthesis, potentially reducing photosynthetic output by 25% in C_3 plants. Photorespiration involves a complex network of enzyme reactions that exchange metabolites between chloroplasts, leaf peroxisomes and mitochondria.

The oxygenation reaction of RuBisCO is a wasteful process because 3-phosphoglycerate is created at a reduced rate and higher metabolic cost compared with RuBP carboxylase activity. While photorespiratory carbon cycling results in the formation of G3P eventually, there is still a net loss of carbon (around 25% of carbon fixed by photosynthesis is re-released as CO_2) and nitrogen, as ammonia. Ammonia must be detoxified at a substantial cost to the cell. Photorespiration also incurs a direct cost of one ATP and one NAD(P)H.

While it is common to refer to the entire process as photorespiration, technically the term refers only to the metabolic network which acts to rescue the products of the oxygenation reaction (phosphoglycolate).

Photorespiratory Reactions

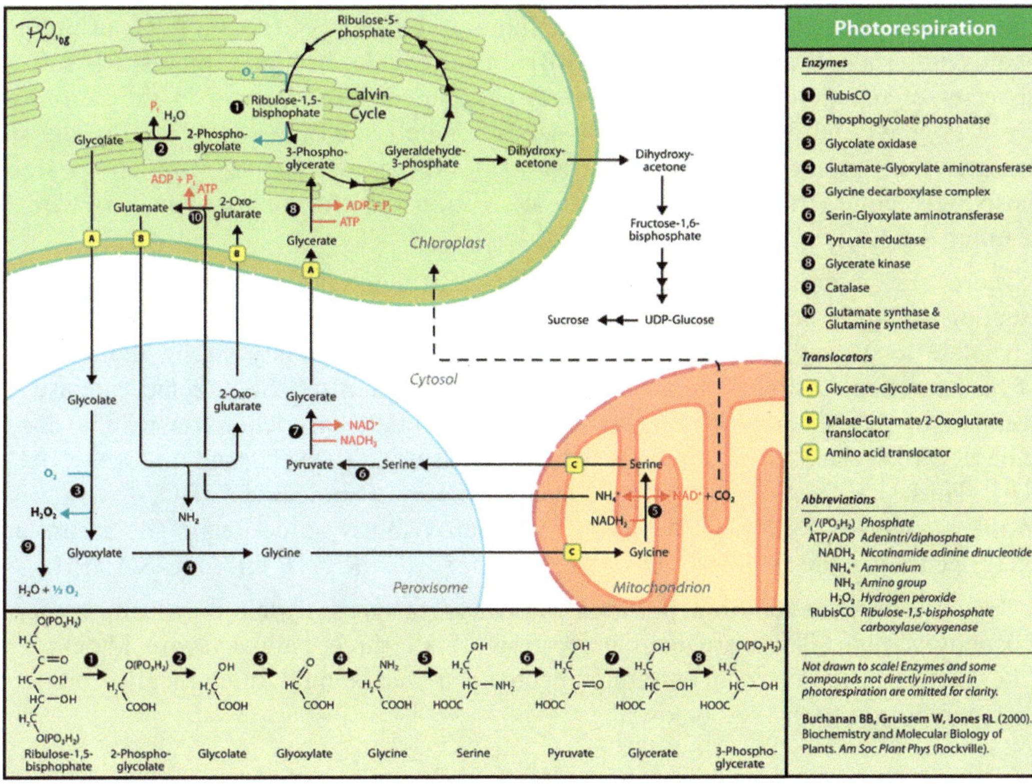

Photorespiration

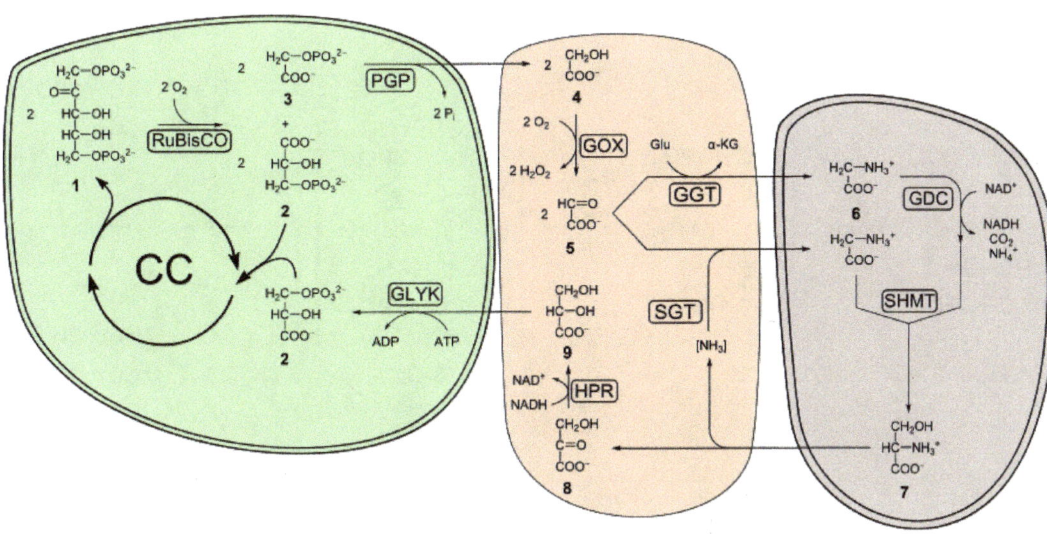

Photorespiration

Addition of molecular oxygen to ribulose-1,5-bisphosphate produces 3-phosphoglycerate (PGA) and 2-phosphoglycolate (2PG, or PG). PGA is the normal product of carboxylation, and productively enters the Calvin cycle. Phosphoglycolate, however, inhibits certain enzymes involved in photosynthetic carbon fixation (hence is often said to be an 'inhibitor of photosynthesis'). It is also relatively difficult to recycle: in higher plants it is salvaged by a series of reactions in the peroxisome, mitochondria, and again in the peroxisome where it is converted into glycerate. Glycerate reenters the chloroplast and by the same transporter that exports glycolate. A cost of 1 ATP is associated with conversion to 3-phosphoglycerate (PGA) (Phosphorylation), within the chloroplast, which is then free to re-enter the Calvin cycle.

There are several costs associated with this metabolic pathway; one being the production of hydrogen peroxide in the peroxisome (associated with the conversion of glycolate to glyoxylate). Hydrogen peroxide is a dangerously strong oxidant which must be immediately broken down into water and oxygen by the enzyme catalase. The conversion of 2x 2Carbon glycine to 1 C3 serine in the mitochondria by the enzyme glycine-decarboxylase is a key step, which releases CO_2, NH_3, and reduces NAD to NADH. Thus, 1 CO_2 molecule is produced for every 3 molecules of O_2 (two deriving from the activity of RuBisCO, the third from peroxisomal oxidations). The assimilation of NH_3 occurs via the GS-GOGAT cycle, at a cost of one ATP and one NADPH.

Cyanobacteria have three possible pathways through which they can metabolise 2-phosphoglycolate. They are unable to grow if all three pathways are knocked out, despite having a carbon concentrating mechanism that should dramatically reduce the rate of photorespiration (see below).

Substrate Specificity of Rubisco

The oxidative photosynthetic carbon cycle reaction is catalyzed by RuBP oxygenase activity:

$$RuBP + O_2 \rightarrow \text{Phosphoglycolate} + \text{3-phosphoglycerate} + 2H^+$$

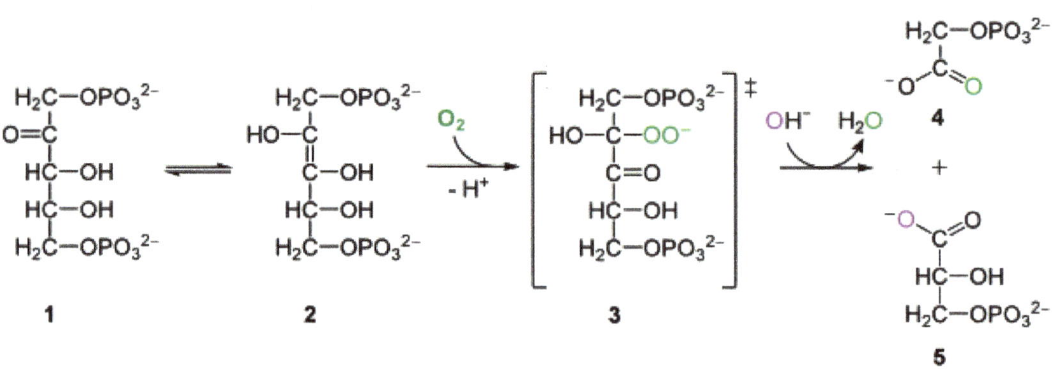

Oxygenase activity of RuBisCO

During the catalysis by RuBisCO, an 'activated' intermediate is formed (an enediol intermediate) in the RuBisCO active site. This intermediate is able to react with either CO_2 or O_2. It has been demonstrated that the specific shape of the RuBisCO active site acts to encourage reactions with CO_2. Although there is a significant "failure" rate (~25% of reactions are oxygenation rather than carboxylation), this represents significant favouring of CO_2, when the relative abundance of the two gasses is taken into account: in the current atmosphere, O_2 is approximately 500 times more abundant, and in solution O_2 is 25X more abundant than CO_2. The ability of RuBisCO to specify between the two gasses is known as its selectivity factor (or Srel), and it varies between species, with angiosperms more efficient than other plants, but with little variation among the vascular plants.

A suggested explanation into RuBisCO's inability to discriminate completely between CO_2 and O_2 is that it is an evolutionary relic: The early atmosphere in which primitive plants originated contained very little oxygen, the early evolution of RuBisCO was not influenced by its ability to discriminate between O_2 and CO_2.

Conditions which Increase Photorespiration

Photorespiration rates are increased by:

Altered substrate availability: Lowered CO_2 or Increased O_2

Factors which influence this include the atmospheric abundance of the two gases, the supply of the gases to the site of fixation (i.e. in land plants: whether the stomata are open or closed), the length of the liquid phase (how far these gases have to diffuse through water in order to reach the reaction site). For example, when the stomata are closed to prevent water loss during drought: this limits the CO_2 supply, while O_2 production within the leaf will continue. In algae (and plants which photosynthesise underwater); gases have to diffuse significant distances through water, which results in a decrease in the availability of CO_2 relative to O_2. It has been predicted that the increase in ambient CO_2 concentrations predicted over the next 100 years may reduce the rate of photorespiration in most plants by around 50%.

Increased Temperature

At higher temperatures RuBisCO is less able to discriminate between CO_2 and O_2. This is because the enediol intermediate is less stable. Increasing temperatures also reduce the solubility of CO_2, thus reducing the concentration of CO_2 relative to O_2 in the chloroplast.

Biological Adaptation to Minimize Photorespiration

Certain species of plants or algae have mechanisms to reduce uptake of molecular oxygen by RuBisCO. These are commonly referred to as Carbon Concentrating Mechanisms (CCMs), as they increase the concentration of CO_2 so that RuBisCO is less likely to produce glycolate through reaction with O_2.

Biochemical Carbon Concentrating Mechanisms

Biochemical CCMs concentrate carbon dioxide in one temporal or spatial region, through metabolite exchange. C_4 and CAM photosynthesis both use the enzyme Phosphoenolpyruvate carboxylase (PEPC) to add CO_2 to a 3-Carbon sugar. PEPC is faster than RuBisCO, and more selective for CO_2.

C_4

Corn uses the C_4 pathway, minimizing photorespiration.

C_4 plants capture carbon dioxide in their mesophyll cells (using an enzyme called Phosphoenolpyruvate carboxylase which catalyzes the combination of carbon dioxide with a compound called Phosphoenolpyruvate (PEP)), forming oxaloacetate. This oxaloacetate is then converted to malate and is transported into the bundle sheath cells (site of carbon dioxide fixation by RuBisCO) where oxygen concentration is low to avoid photorespiration. Here, carbon dioxide is removed from the malate and combined with RuBP by RuBisCO in the usual way, and the Calvin cycle proceeds as normal. The CO_2 concentrations in the Bundle Sheath are approximately 10-20 fold higher than the concentration in the mesophyll cells.

This ability to avoid photorespiration makes these plants more hardy than other plants in dry and hot environments, wherein stomata are closed and internal carbon dioxide levels are low. Under these conditions, photorespiration does occur in C_4plants, but at a much reduced level compared with C_3 plants in the same conditions. C_4 plants include sugar cane, corn (maize), and sorghum.

CAM (Crassulacean Acid Metabolism)

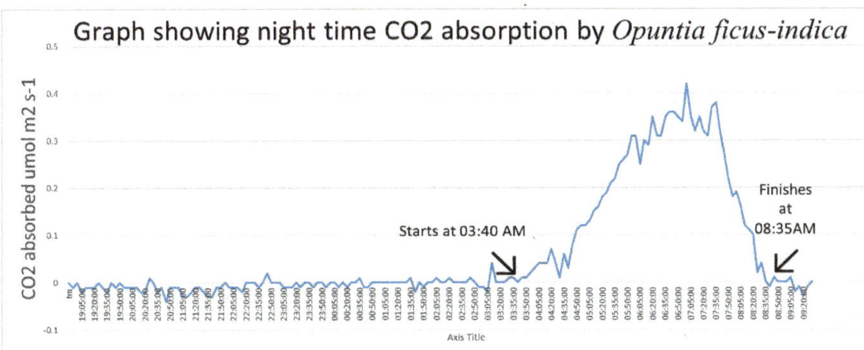

Overnight graph of CO_2 absorbed by a CAM plant

CAM plants, such as cacti and succulent plants, also use the enzyme PEP carboxylase to capture carbon dioxide, but only at night. Crassulacean acid metabolism allows plants to conduct most of their gas exchange in the cooler night-time air, sequestering carbon in 4 carbon sugars which can be released to the photosynthesizing cells during the day. This allows CAM plants to reduce water loss (transpiration) by maintaining closed stomata during the day. CAM plants usually display other water-saving characteristics, such as thick cuticles, stomata with small apertures, and typically lose around 1/3 of the amount of water per CO_2 fixed.

Algae

There have been some reports of algae operating a biochemical CCM: shuttling metabolites within single cells to concentrate CO_2 in one area. This process is not fully understood.

Biophysical Carbon-Concentrating Mechanisms

This type of carbon-concentrating mechanism (CCM) relies on a contained compartment within the cell into which CO_2 is shuttled, and where RuBisCO is highly expressed. In many species, biophysical CCMs are only induced under low carbon dioxide concentrations. Biophysical CCMs are more evolutionarily ancient than biochemical CCMs. There is some debate as to when biophysical CCMs first evolved, but it is likely to have been during a period of low carbon dioxide, after the Great Oxygenation Event (2.4 billion years ago). Low CO_2 periods occurred around 750, 650, and 320-270 million years ago.

Eukaryotic Algae

In nearly all species of eukaryotic algae (*Chloromonas* being one notable exception), upon induction of the CCM, ~95% of RuBisCO is densely packed into a single subcellular compartment: the pyrenoid. Carbon dioxide is concentrated in this compartment using a combination of CO_2 pumps, bicarbonate pumps, and carbonic anhydrases. The pyrenoid is not a membrane bound compartment, but is found within the chloroplast, often surrounded by a starch sheath (which is not thought to serve a function in the CCM).

Hornworts

Certain species of hornwort are the only land plants which are known to have a biophysical CCM involving concentration of carbon dioxide within pyrenoids in their chloroplasts.

Cyanobacteria

Cyanobacterial CCMs are similar in principle to those found in eukaryotic algae and hornworts, but the compartment into which carbon dioxide is concentrated has several structural differences. Instead of the pyrenoid, cyanobacteria contain carboxysomes, which have a protein shell, and linker proteins packing RuBisCO inside with a very regular structure. Cyanobacterial CCMs are much better understood than those found in eukaryotes, partly due to the ease of genetic manipulation of prokaryotes.

Possible Purpose of Photorespiration

Reducing photorespiration may not result in increased growth rates for plants. Photorespiration may be necessary for the assimilation of nitrate from soil. Thus, a reduction in photorespiration by genetic engineering or because of increasing atmospheric carbon dioxide due to fossil fuel burning may not benefit plants as has been proposed. Several physiological processes may be responsible for linking photorespiration and nitrogen assimilation. Photorespiration increases availability of NADH, which is required for the conversion of nitrate to nitrite. Certain nitrite transporters also transport bicarbonate, and elevated CO_2 has been shown to suppress nitrite transport into chloroplasts.

Although photorespiration is greatly reduced in C_4 species, it is still an essential pathway - mutants without functioning 2-phosphoglycolate metabolism cannot grow in normal conditions. One mutant was shown to rapidly accumulate glycolate.

Although the functions of photorespiration remain controversial, it is widely accepted that this pathway influences a wide range of processes from bioenergetics, photosystem II function, and carbon metabolism to nitrogen assimilation and respiration. The photorespiratory pathway is a major source of hydrogen peroxide (H_2O_2) in photosynthetic cells. Through H_2O_2 production and pyridine nucleotide interactions, photorespiration makes a key contribution to cellular redox homeostasis. In so doing, it influences multiple signalling pathways, in particular, those that govern plant hormonal responses controlling growth, environmental and defense responses, and programmed cell death.

Another theory postulates that it may function as a «safety valve», preventing the excess of reductive potential coming from an overreduced NADPH-pool from reacting with oxygen and producing free radicals, as these can damage the metabolic functions of the cell by subsequent oxidation of membrane lipids, proteins or nucleotides.

REFERENCES

1. Geider, R. J., et al., "Primary productivity of planet earth: biological determinants and physical constraints in terrestrial and aquatic habitats", Global Change Biol. 2001, 7, 849–882. doi:10.1046/j.1365-2486.2001.00448.x
2. Swan BK, Martinez-Garcia M, Preston CM, Sczyrba A, Woyke T, Lamy D, Reinthaler T, Poulton NJ, Masland ED, Gomez ML, Sieracki ME, DeLong EF, Herndl GJ, Stepanauskas R (2011). "Potential for chemolithoautotrophy among ubiquitous bacteria lineages in the dark ocean". Science. 333 (6047): 1296–300. Bibcode:2011Sci...333.1296S. doi:10.1126/science.1203690. PMID 21885783.
3. Brasier M, McLoughlin N, Green O, Wacey D (2006). "A fresh look at the fossil evidence for early Archaean cellular life". Philos. Trans. R. Soc. Lond., B, Biol. Sci. 361 (1470): 887–902. doi:10.1098/rstb.2006.1835. PMC 1578727 . PMID 16754605.
4. Tomitani A, Knoll AH, Cavanaugh CM, Ohno T (2006). "The evolutionary diversification of cyanobacteria: molecular-phylogenetic and paleontological perspectives". Proc. Natl. Acad. Sci. U.S.A. 103 (14): 5442–7. Bibcode:2006PNAS..103.5442T. doi:10.1073/pnas.0600999103. PMC 1459374 . PMID 16569695.
5. Kopp RE, Kirschvink JL, Hilburn IA, Nash CZ (2005). "The Paleoproterozoic snowball Earth: a climate disaster triggered by the evolution of oxygenic photosynthesis". Proc. Natl. Acad. Sci. U.S.A. 102 (32): 11131–6. Bibcode:2005PNAS..10211131K. doi:10.1073/pnas.0504878102. PMC 1183582 . PMID 16061801.
6. Dodd AN, Borland AM, Haslam RP, Griffiths H, Maxwell K (2002). "Crassulacean acid metabolism: plastic, fantastic". J. Exp. Bot. 53 (369): 569–580. doi:10.1093/jexbot/53.369.569. PMID 11886877.
7. O'Leary MH (1988). "Carbon isotopes in photosynthesis". BioScience. 38 (5): 328–336. doi:10.2307/1310735. JSTOR 1310735.
8. Sage RF, Meirong L, Monson RK (1999). "16. The Taxonomic Distribution of C4 Photosynthesis". In Sage RF, Monson RK. C4 Plant Biology. pp. 551–580. ISBN 0-12-614440-0.
9. Evans MC, Buchanan BB, Arnon DI (1966). "A new ferredoxin-dependent carbon reduction cycle in a photosynthetic bacterium". Proc. Natl. Acad. Sci. U.S.A. 55 (4): 928–34. Bibcode:1966PNAS...55..928E. doi:10.1073/pnas.55.4.928. PMC 224252 . PMID 5219700.
10. Ljungdahl L, Wood HG (1965). "Incorporation of C-14 from carbon dioxide into sugar phosphates, carboxylic acids, and amino acids by Clostridium thermoaceticum". J. Bacteriol. 89: 1055–64. PMC 277595 . PMID 14276095.
11. Ljungdahl LG (2009). "A life with acetogens, thermophiles, and cellulolytic anaerobes". Annu. Rev. Microbiol. 63: 1–25. doi:10.1146/annurev.micro.091208.073617. PMID 19575555.

12. Herter S, Fuchs G, Bacher A, Eisenreich W (2002). "A bicyclic autotrophic CO2fixation pathway in Chloroflexus aurantiacus". J. Biol. Chem. 277 (23): 20277–83. doi:10.1074/jbc.M201030200. PMID 11929869.

13. Zarzycki J, Brecht V, Müller M, Fuchs G (2009). "Identifying the missing steps of the autotrophic 3-hydroxypropionate CO2 fixation cycle in Chloroflexus aurantiacus". Proc. Natl. Acad. Sci. U.S.A. 106 (50): 21317–22. Bibcode:2009PNAS..10621317Z. doi:10.1073/pnas.0908356106. PMC 2795484 . PMID 19955419.

14. Berg IA, Kockelkorn D, Buckel W, Fuchs G (2007). "A 3-hydroxypropionate/4-hydroxybutyrate autotrophic carbon dioxide assimilation pathway in Archaea". Science. 318 (5857): 1782–6. Bibcode:2007Sci...318.1782B. doi:10.1126/science.1149976. PMID 18079405.

15. Huber H, Gallenberger M, Jahn U, Eylert E, Berg IA, Kockelkorn D, Eisenreich W, Fuchs G (2008). "A dicarboxylate/4-hydroxybutyrate autotrophic carbon assimilation cycle in the hyperthermophilic Archaeum Ignicoccus hospitalis". Proc. Natl. Acad. Sci. U.S.A. 105 (22): 7851–6. Bibcode:2008PNAS..105.7851H. doi:10.1073/pnas.0801043105. PMC 2409403 . PMID 18511565.

16. Encyclopedia of Microbiology. Academic Press. 2009. pp. 83–84. ISBN 9780123739445.

17. Nicole Kresge; Robert D. Simoni; Robert L. Hill (2005). "The Discovery of Heterotrophic Carbon Dioxide Fixation by Harland G. Wood". The Journal of Biological Chemistry.

CHAPTER 7

Crassulacean Acid Metabolism CAM Photosynthesis

Chapter Contents

Crassulacean acid metabolism (CAM) is a photosynthetic CO2 fixation pathway that maximizes water use efficiency (WUE) many times relative to C3 species by using a temporal CO2 pump. Thus, CAM provides an excellent opportunity to engineer both enhanced photosynthetic performance and WUE into bioenergy crops. The proposed research will provide a comprehensive understanding of the enzymatic and regulatory pathways required to engineer CAM photosynthetic machinery into Populus. The methods employed will include deep transcriptome sequencing (RNA-Seq) and high-throughput metabolic profiling of leaf mesophyll cells and stomatal guard cells to identify regulators of the nocturnal opening and daytime closure of stomata that underpins the high WUE of CAM plants in order to create co-expression models.

INTRODUCTION

Crassulacean acid metabolism, also known as CAM photosynthesis, is a carbon fixation pathway that evolved in some plants as an adaptation to arid conditions. In a plant using full CAM, the stomata in the leaves remain shut during the day to reduce evapotranspiration, but open at night to collect carbon dioxide (CO2). The CO2 is stored as the four-carbon acid malate in vacuoles at night, and then in the daytime, the malate is transported to chloroplasts where it is converted back to CO2, which is then used during photosynthesis. The pre-collected CO2 is concentrated around the enzyme RuBisCO, increasing photosynthetic efficiency. This metabolism was first studied in plants of the Crassulaceae family. These mainly include succulents. The first time it was studied, Crassula was used as a model organism.

The pineapple is an example of a CAM plant.

HISTORICAL BACKGROUND

CAM was first suspected by de Saussure in 1804 in his *Recherches Chimiques sur la Vegetation*, confirmed and refined by Aubert, E. in 1892 in his *Recherches physiologiques sur les plantes grasses* and expounded upon by Richards, H. M. 1915 in *Acidity and Gas Interchange in Cacti*, Carnegie Institution. The term CAM may have been coined by Ranson and Thomas in 1940, but they were not the first to discover this cycle. It was observed by the botanists Ranson and Thomas, in the Crassulaceae family of succulents (which includes jade plants and *Sedum*). Its name refers to acid metabolism in Crassulaceae, not the metabolism of "crassulacean acid".

Overview: A Two-Part Cycle

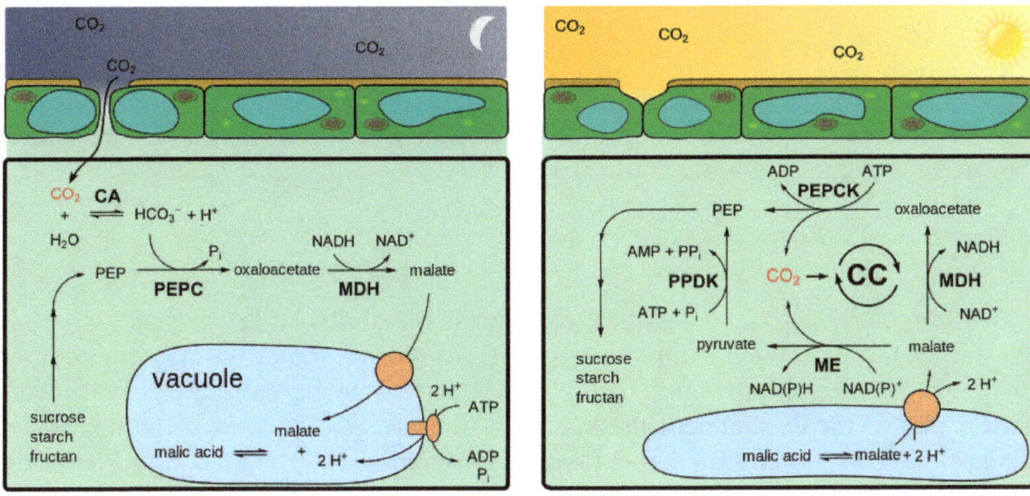

Overview of CAM

CAM is an adaptation for increased efficiency in the use of water, and so is typically found in plants growing in arid conditions.

During the Night

During the night, a plant employing CAM has its stomata open, allowing CO_2 to enter and be fixed as organic acids that are stored in vacuoles. During the day the stomata are closed (thus preventing water loss), and the carbon is released to the Calvin cycle so that photosynthesis may take place.

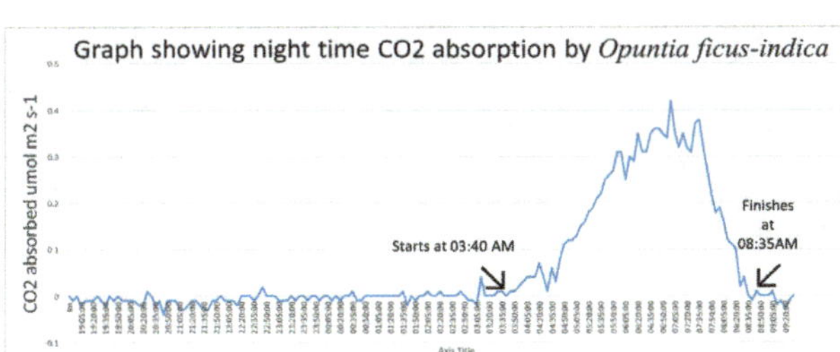

Overnight graph of CO2 absorbed by a CAM plant

The carbon dioxide is fixed in the cytoplasm of mesophyll cells by a PEP reaction similar to that of C_4 pathway. But, unlike the C_4 mechanism, the resulting organic acids are stored in vacuoles for later use; that is, they are not immediately passed on to the

Calvin cycle. The latter cannot operate during the night because the light reactions that provide it with ATP and NADPH cannot take place.

During the Day

During the day, the CO_2-storing organic acids are released from the vacuoles of the mesophyll cells and enter the stroma of the chloroplasts where an enzyme releases the CO_2, which then enters into the Calvin cycle.

Benefits

The most important benefit of CAM to the plant is the ability to leave most leaf stomata closed during the day. Plants employing CAM are most common in arid environments, where water comes at a premium. Being able to keep stomata closed during the hottest and driest part of the day reduces the loss of water through evapotranspiration, allowing such plants to grow in environments that would otherwise be far too dry. Plants using only C_3 carbon fixation, for example, lose 97% of the water they uptake through the roots to transpiration - a high cost avoided by plants able to employ CAM.

Comparison with C_4 Metabolism

CAM is named after the family Crassulaceae, to which Jade plant belongs

The C_4 pathway bears resemblance to CAM; both act to concentrate CO_2 around RuBisCO, thereby increasing its efficiency. CAM concentrates it temporally, providing CO_2 during the day, and not at night, when respiration is the dominant reaction. C_4plants, in contrast, concentrate CO_2 spatially, with a RuBisCO reaction centre in a "bundle sheath cell" being inundated with CO_2. Due to the inactivity required by the CAM mechanism, C_4 carbon fixation has a greater efficiency in terms of PGA synthesis

Biochemistry

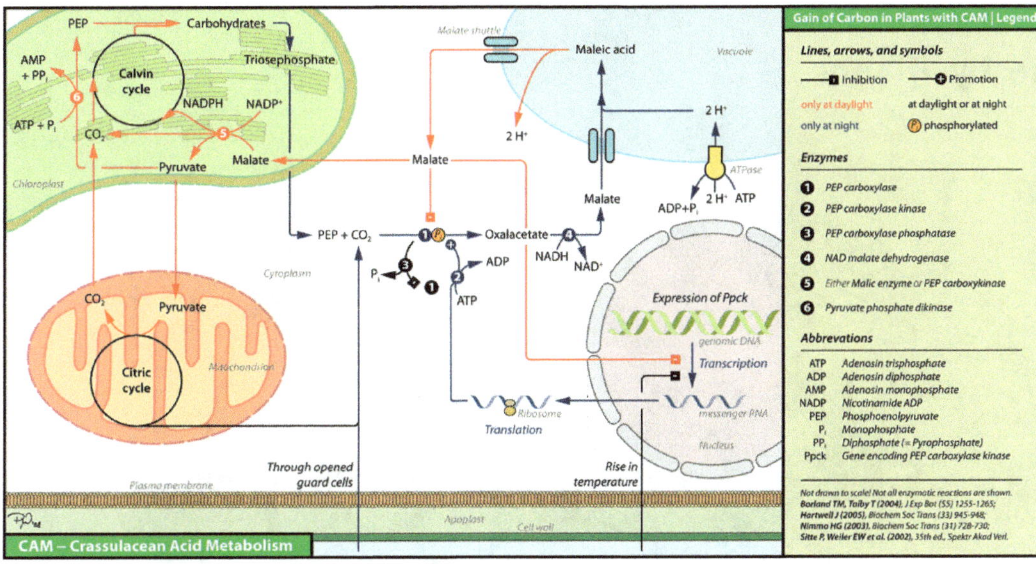

Biochemistry of CAM

Plants with CAM must control storage of CO_2 and its reduction to branched carbohydrates in space and time.

At low temperatures (frequently at night), plants using CAM open their stomata, CO_2 molecules diffuse into the spongy mesophyll's intracellular spaces and then into the cytoplasm. Here, they can meet phosphoenolpyruvate (PEP), which is a phosphorylated triose. During this time, the plants are synthesizing a protein called PEP carboxylase kinase (PEP-C kinase), whose expression can be inhibited by high temperatures (frequently at daylight) and the presence of malate. PEP-C kinase phosphorylates its target enzyme PEP carboxylase (PEP-C). Phosphorylation dramatically enhances the enzyme›s capability to catalyze the formation of oxalacetate, which can be subsequently transformed into malate by NAD^+malate dehydrogenase. Malate is then transported via malate shuttles into the vacuole, where it is converted into the storage form malic acid. In contrast to PEP-C kinase, PEP-C is synthesized all the time but almost inhibited at daylight either by dephosphorylation via PEP-C phosphatase or directly by binding malate. The latter is not possible at low temperatures, since malate is efficiently transported into the vacuole, whereas PEP-C kinase readily inverts dephosphorylation.

In daylight, plants using CAM close their guard cells and discharge malate that is subsequently transported into chloroplasts. There, depending on plant species, it is cleaved into pyruvate and CO_2 either by malic enzyme or by PEP carboxykinase. CO_2 is then introduced into the Calvin cycle, a coupled and self-recovering enzyme system, which is used to build branched carbohydrates. The by-product pyruvate can be further degraded in the mitochondrial citric acid cycle, thereby providing additional CO_2 molecules for the Calvin Cycle. Pyruvate can also be used to recover PEP via pyruvate phosphate dikinase, a high-energy step, which requires ATP and an additional phosphate. During the following cool night, PEP is finally exported into the cytoplasm, where it is involved in fixing carbon dioxide via malate.

Use by Plants

Cross section of a CAM (Crassulacean acid metabolism) plant, specifically of an agave leaf. Vascular bundles shown. Drawing based on microscopic images courtesy of Cambridge University Plant Sciences Department.

Plants use CAM to different degrees. Some are "obligate CAM plants", i.e. they use only CAM in photosynthesis, although they vary in the amount of CO_2 they are able to store as organic acids; they are sometimes divided into "strong CAM" and "weak CAM" plants on this basis. Other plants show "inducible CAM", in which they are able to switch between using either the C_3 or C_4mechanism and CAM depending on environmental conditions. Another group of plants employ "CAM-cycling", in which their stomata do not open at night; the plants instead recycle CO_2produced by respiration as well as storing some CO_2 during the day.

Plants showing inducible CAM and CAM-cycling are typically found in conditions where periods of water shortage alternate with periods when water is freely available. Periodic drought – a feature of semi-arid regions – is one cause of water shortage. Plants which grow on trees or rocks (as epiphytes or lithophytes) also experience variations in water availability. Salinity, high light levels and nutrient availability are other factors which have been shown to induce CAM.

Since CAM is an adaptation to arid conditions, plants using CAM often display other xerophytic characters, such as thick, reduced leaves with a low surface-area-to-volume ratio; thick cuticle; and stomata sunken into pits. Some shed their leaves during the dry season; others (the succulents) store water in vacuoles. CAM also causes taste differences: plants may have an increasingly sour taste during the night yet become sweeter-tasting during the day. This is due to malic acid being stored in the vacuoles of the plants› cells during the night and then being used up during the day.

AQUATIC CAM

CAM photosynthesis is also found in aquatic species in at least 4 genera, including: *Isoetes*, *Crassula*, *Littorella*, *Sagittaria*, and possibly *Vallisneria*, being found in a variety of species e.g. *Isoetes howellii*, *Crassula aquatica*.

These plants follow the same nocturnal acid accumulation and daytime deacidification as terrestrial CAM species. However, the reason for CAM in aquatic plants is not due to a lack of available water, but a limited supply of CO_2. CO_2 is limited due to slow diffusion in water, 10000x slower than in air. The problem is especially acute under acid pH, where the only inorganic carbon species present is CO_2, with no available bicarbonate or carbonate supply.

Aquatic CAM plants capture carbon at night when it is abundant due to a lack of competition from other photosynthetic organisms. This also results in lowered *photorespiration* due to less photosynthetically generated oxygen.

Aquatic CAM is most marked in the summer months when there is increased competition for CO_2, compared to the winter months. However, in the winter months CAM still has a significant role.

Ecological and Taxonomic Distribution of CAM-Using Plants

The majority of plants possessing CAM are either epiphytes (e.g., orchids, bromeliads) or succulent xerophytes (e.g., cacti, cactoid *Euphorbias*), but CAM is also found in hemiepiphytes (e.g., *Clusia*); lithophytes (e.g., *Sedum*, *Sempervivum*); terrestrial bromeliads; wetland plants (e.g., *Isoetes*, *Crassula* (*Tillaea*), *Lobelia*;and in one halophyte, *Mesembryanthemum crystallinum*; one non-succulent terrestrial plant, (*Dodonaea viscosa*) and one mangrove associate (*Sesuvium portulacastrum*).

Plants which are able to switch between different methods of carbon fixation include *Portulacaria afra*, better known as Dwarf Jade Plant, which normally uses C_3fixation but can use CAM if it is drought-stressed, and *Portulaca oleracea*, better known as Purslane, which normally uses C_4 fixation but is also able to switch to CAM when drought-stressed.

CAM has evolved convergently many times. It occurs in 16,000 species (about 7% of plants), belonging to over 300 genera and around 40 families, but this is thought to be a considerable underestimate. It is found in quillworts (relatives of club mosses), in ferns, and in Gnetopsida, but the great majority of plants using CAM are angiosperms (flowering plants).

The following list summarizes the taxonomic distribution of plants with CAM.

Division	Class/Angiosperm group	Order	Family	Plant Type	Clade involved
Lycopodiophyta	Isoetopsida	Isoetales	Isoetaceae	hydrophyte	*Isoetes* (the sole genus of class Isoetopsida) - *I. howellii* (seasonally submerged), *I. macrospora*, *I. bolanderi*, *I. engelmanni*, *I. lacustris*, *I. sinensis*, *I. storkii*, *I. kirkii*
Pteridophyta	Polypodiopsida	Polypodiales	Polypodiaceae	epiphyte, lithophyte	CAM is recorded from *Microsorium*, *Platycerium* and *Polypodium*, *Pyrrosia* and *Drymoglossum* and *Microgramma*
	Pteridopsida	Polypodiales	Pteridaceae	epiphyte	*Vittaria* *Anetium citrifolium*
Cycadophyta	Cycadopsida	Cycadales	Zamiaceae		*Dioon edule*
Pinophyta	Gnetopsida	Welwitschiales	Welwitschiaceae	xerophyte	*Welwitschia mirabilis* (the sole species of the order Welwitschiales)
Magnoliophyta	magnoliids	Magnoliales	Piperaceae	epiphyte	*Peperomia camptotricha*
	eudicots	Caryophyllales	Aizoaceae	xerophyte	widespread in the family; *Mesembryanthemum crystallinum* is a rare instance of an halophyte that displays CAM
			Cactaceae	xerophyte	Almost all cacti have obligate Crassulacean Acid Metabolism in their stems; the few cacti with leaves may have C_3 Metabolism in those leaves; seedlings have C_3 Metabolism.
			Portulacaceae	xerophyte	recorded in approximately half of the genera (note: Portulacaceae is paraphyletic with respect to Cactaceae and Didieraceae)
			Didiereaceae	xerophyte	
		Saxifragales	Crassulaceae	hydrophyte, xerophyte, lithophyte	CAM is widespread in the family
	eudicots (rosids)	Vitales	Vitaceae		*Cissus*, *Cyphostemma*
		Malpighiales	Clusiaceae	hemiepiphyte	*Clusia*
			Euphorbiaceae		CAM is found is some species of *Euphorbia* including some formerly placed in the sunk genera *Monadenium*,*Pedilanthus* and *Synadenium*. C4 photosynthesis is also found in *Euphorbia* (subgenus *Chamaesyce*).
			Passifloraceae	xerophyte	*Adenia*
		Geraniales	Geraniaceae		CAM is found in some succulent species of *Pelargonium*,and is also reported from *Geranium pratense*
		Cucurbitales	Cucurbitaceae		*Xerosicyos danguyi*, *Dendrosicyos socotrana*, *Momordica*
		Celastrales	Celastraceae		
		Oxalidales	Oxalidaceae		*Oxalis carnosa* var. *hirta*
		Brassicales	Moringaceae		*Moringa*
			Salvadoraceae		CAM is found in *Salvadora persica*. Salvadoraceae were previously placed in order Celastrales, but are now placed in Brassicales.
		Sapindales	Sapindaceae		*Dodonaea viscosa*

		Fabales	Fabaceae		CAM is found in *Prosopis juliflora* (listed under the family Salvadoraceae in Sayed's (2001) table,) but is currently in the family Fabaceae (Leguminosae) according to The Plant List).
			Zygophyllaceae		*Zygophyllum*
	eudicots (asterids)	Ericales	Ebenaceae		
		Solanales	Convolvulaceae		*Ipomoea* (Some species of Ipomoea are C3- a citation is needed here.)
		Gentianales	Rubiaceae	epiphyte	*Hydnophytum* and *Myrmecodia*
			Apocynaceae		CAM is found in subfamily Asclepidioideae *(Hoya*, *Dischidia*, *Ceropegia*, *Stapelia*, *Caralluma negevensis*, *Frerea indica*, *Adenium*, *Huernia*), and also in *Carissa*and *Akocanthera*
		Lamiales	Gesneriaceae	epiphyte	CAM was found *Codonanthe crassifolia*, but not in 3 other genera
			Lamiaceae		*Plectranthus marrubioides*, *Coleus*
			Plantaginaceae	hydrophyte	*Littorella uniflora*
		Apiales	Apiaceae	hydrophyte	*Lilaeopsis lacustris*
		Asterales	Asteraceae		some species of *Senecio*
Magnoliophyta	monocots	Alismatales	Hydrocharitaceae	hydrophyte	*Hydrilla*, *Vallisneria*
			Alismataceae	hydrophyte	*Sagittaria*
			Araceae		*Zamioculcas zamiifolia* is the only CAM plant in Araceae, and the only non-aquatic CAM plant in Alismatales
		Poales	Bromeliaceae	epiphyte	Bromelioideae (91%), *Puya* (24%), *Dyckia* and related genera (all), *Hechtia* (all), *Tillandsia* (many)
			Cyperaceae	hydrophyte	*Scirpus*, *Eleocharis*
		Asparagales	Orchidaceae	epiphyte	The Orchidaceae has more CAM species than any other family (CAM Orchids)
			Agavaceae	xerophyte	*Agave*, *Hesperaloe*, *Yucca* and *Polianthes*
			Asphodelaceae	xerophyte	*Aloe*, *Gasteria*, and *Haworthia*
			Ruscaceae		*Sansevieria* (This genus is listed under the family Dracaenaceae in Sayed's (2001) table, but currently in the family Asparagaceae according to The Plant List), *Dracaena*
		Commelinales	Commelinaceae		*Callisia*, *Tradescantia*, *Tripogandra*

REFERENCES

1. C.Michael Hogan. 2011. Respiration. Encyclopedia of Earth. Eds. Mark McGinley & C.J.cleveland. National council for Science and the Environment. Washington DC
2. Ranson S. L.; Thomas M (1960). "Crassulacean acid metabolism". Annu Rev Plant Physiol. 11 (1): 81–110. doi:10.1146/annurev.pp.11.060160.000501.
3. Herrera, A. (2008), "Crassulacean acid metabolism and fitness under water deficit stress: if not for carbon gain, what is facultative CAM good for?", Annals of Botany, 103 (4): 645–653, doi:10.1093/aob/mcn145
4. Ting, I P (1985). "Crassulacean Acid Metabolism". Annual Review of Plant Physiology. 36 (1): 595–622. doi:10.1146/annurev.pp.36.060185.003115.
5. Raven JA, Edwards D (2001). "Roots: evolutionary origins and biogeochemical significance". Journal of Experimental Botany. 52 (90001): 381–401. doi:10.1093/jexbot/52.suppl_1.381. PMID 11326045.
6. Raven, P & Evert, R & Eichhorn, S, 2005, "Biology of Plants" (seventh edition), p. 135 (Figure 7-26), W.H. Freeman and Company Publishers ISBN 0-7167-1007-2
7. Keeley, J (1998). "CAM Photosynthesis in Submerged Aquatic Plants". Botanical Review. 64 (2): 121–175. doi:10.1007/bf02856581.
8. Keeley, J; Busch, G (1984). "Carbon Assimilation Characteristics of the Aquatic CAM Plant, Isoetes howellii". Plant Physiology. 75: 525–530. doi:10.1104/pp.76.2.525.
9. Klavsen, S; Madsen, T (2012). "Seasonal variation in crassulacean acid metabolism by the aquatic isoetid Littorella uniflora". Photosynthesis Research. 112: 163–173. doi:10.1007/s11120-012-9759-0.
10. Keddy, P.A. (2010). Wetland Ecology: Principles and Conservation. Cambridge University Press, Cambridge, UK. p. 26.
11. Guralnick, L. J.; Ting, I. P. (1987). "Physiological Changes in Portulacaria afra (L.) Jacq. during a Summer Drought and Rewatering". Plant Physiology. 85 (2): 481–6. doi:10.1104/pp.85.2.481. PMC 1054282 . PMID 16665724.
12. Koch, K. E.; Kennedy, R. A. (1982). "Crassulacean Acid Metabolism in the Succulent C4 Dicot, Portulaca oleracea L Under Natural Environmental Conditions". Plant Physiology. 69 (4): 757–61. doi:10.1104/pp.69.4.757. PMC 426300 . PMID 16662291.
13. Keeley, Jon E.; Rundel, Philip W. (2003). "Evolution of CAM and C4 Carbon-Concentrating Mechanisms" (PDF). International Journal of Plant Sciences. 164 (S3): S55. doi:10.1086/374192.
14. Dodd, A. N.; Borland, A. M.; Haslam, R. P.; Griffiths, H.; Maxwell, K. (2002). "Crassulacean acid metabolism: plastic, fantastic". Journal of Experimental Botany. 53 (369): 569–580. doi:10.1093/jexbot/53.369.569. PMID 11886877.
15. Boston, H (1983). "Evidence of crussulacean acid metabolism in two North American isoetids". Aquatic Botany. 15 (4): 381–386. doi:10.1016/0304-

3770(83)90006-2.

16. Holtum, Joseph A.M.; Winter, Klaus (1999). "Degrees of crassulacean acid metabolism in tropical epiphytic and lithophytic ferns". Australian Journal of Plant Physiology. 26 (8): 749. doi:10.1071/PP99001.
17. Wong, S.C.; Hew, C.S (1976). "Diffusive Resistance, Titratable Acidity, and CO2 Fixation in Two Tropical Epiphytic Ferns". American Fern Journal. American Fern Society. 66 (4): 121–124. doi:10.2307/1546463. JSTOR 1546463.
18. Crassulacean Acid Metabolism
19. abstract to Carter & Martin, The occurrence of Crassulacean acid metabolism among ephiphytes in a high-rainfall region of Costa Rica, Selbyana 15(2): 104-106 (1994)
20. Martin, Shannon L.; Davis, Ryan; Protti, Piero; Lin, Teng-Chiu; Lin, Shin-Hwei; Martin, Craig E. (2005). "The Occurrence of Crassulacean Acid Metabolism in Epiphytic Ferns, with an Emphasis on the Vittariaceae". International Journal of Plant Sciences. 166 (4): 623–630. doi:10.1086/430334.
21. Vovides, Andrew P.; Etherington, John R.; Dresser, P. Quentin; Groenhof, Andrew; Iglesias, Carlos; Ramirez, Jonathan Flores (2002). "CAM-cycling in the cycad Dioon edule Lindl. in its natural tropical deciduous forest habitat in central Veracruz, Mexico". Botanical Journal of the Linnean Society. 138 (2): 155–162. doi:10.1046/j.1095-8339.2002.138002155.x.
22. Schulze, E. D.; Ziegler, H.; Stichler, W. (1976). "Environmental control of crassulacean acid metabolism in Welwitschia mirabilis Hook. Fil. in its range of natural distribution in the Namib desert". Oecologia. 24 (4): 323–334. doi:10.1007/BF00381138.
23. Sipes, DL; Ting, IP (1985). "Crassulacean Acid Metabolism and Crassulacean Acid Metabolism Modifications in Peperomia camptotricha.". Plant Physiology. 77 (1): 59–63. doi:10.1104/pp.77.1.59. PMC 1064456 . PMID 16664028.
24. Chu, C; Dai, Z; Ku, MS; Edwards, GE (1990). "Induction of Crassulacean Acid Metabolism in the Facultative Halophyte Mesembryanthemum crystallinum by Abscisic Acid.". Plant Physiology. 93 (3): 1253–1260. doi:10.1104/pp.93.3.1253. PMC 1062660 . PMID 16667587.
25. Nobel, Park S.; Hartsock, Terry L. (1986). "Leaf and Stem CO2 Uptake in the Three Subfamilies of the Cactaceae". Plant Physiology. 80 (4): 913–917. doi:10.1104/pp.80.4.913.
26. Winter, K.; Garcia, M.; Holtum, J. A. M. (2011). "Drought-stress-induced up-regulation of CAM in seedlings of a tropical cactus, Opuntia elatior, operating predominantly in the C3 mode". Journal of Experimental Botany. 62 (11): 4037–4042. doi:10.1093/jxb/err106. ISSN 0022-0957.
27. Guralnick, Lonnie J.; Jackson, Michael D. (2001). "The Occurrence and Phylogenetics of Crassulacean Acid Metabolism in the Portulacaceae". International Journal of Plant Sciences. 162 (2): 257–262. doi:10.1086/319569.
28. Cockburn, W. (1985). "TANSLEY REVIEW No 1.. VARIATION IN PHOTOSYNTHETIC ACID METABOLISM IN VASCULAR PLANTS:

CAM AND RELATED PHENOMENA". New Phytologist. 101 (1): 3–24. doi:10.1111/j.1469-8137.1985.tb02815.x.

29. Nelson, Elizabeth A.; Sage, Tammy L.; Sage, Rowan F. (2005). "Functional leaf anatomy of plants with crassulacean acid metabolism". Functional Plant Biology. 32 (5): 409. doi:10.1071/FP04195.
30. Lüttge, U (2004). "Ecophysiology of Crassulacean Acid Metabolism (CAM).". Annals of Botany. 93 (6): 629–52. doi:10.1093/aob/mch087. PMID 15150072.
31. Bender, MM; Rouhani, I.; Vines, H. M.; Black, C. C. (1973). "C/C Ratio Changes in Crassulacean Acid Metabolism Plants.". Plant Physiology. 52 (5): 427–430. doi:10.1104/pp.52.5.427. PMC 366516 . PMID 16658576.
32. Szarek, S.R. (1979). "The occurrence of Crassulacean Acid Metabolism a supplementary list during 1976 to 1979". Photosynthetica. 13 (4): 467–473.
33. Jones, Cardon & Czaja (2003). "A phylogenetic view of low-level CAM inPelargonium (Geraniaceae)". American Journal of Botany. 90 (1): 135–142. doi:10.3732/ajb.90.1.135.
34. Bastide, Sipes, Hann & Ting (1993). "Effect of Severe Water Stress on Aspects of Crassulacean Acid Metabolism in Xerosicyos.". Plant Physiol. 103 (4): 1089–1096. doi:10.1104/pp.103.4.1089. PMC 159093 . PMID 12232003.
35. Sayed, O.H. (2001). "Crassulacean Acid Metabolism 1975–2000, a Check List". Photosynthetica. 39 (3): 339–352. doi:10.1023/A:1020292623960.
36. "Prosopis juliflora". The Plant List. Retrieved 2015-09-11.
37. Craig E. Martin; Anne E. Lubbers; James A. Teeri (1982). "Variability in Crassulacean Acid Metabolism A Survey of North Carolina Succulent Species". Botanical Gazette. The University of Chicago Press. 143: 491–497. doi:10.1086/337326. JSTOR 2474765.
38. Lange, Otto L.; Zuber, Margit (1977). "Frerea indica, a stem succulent CAM plant with deciduous C3 leaves". Oecologia. 31 (1): 67–72. doi:10.1007/BF00348709.
39. Guralnick; Ting, Irwin P; Lord, Elizabeth M; et al. (1986). "Crassulacean Acid Metabolism in the Gesneriaceae". American Journal of Botany. Botanical Society of America. 73 (3): 336–345. doi:10.2307/2444076. JSTOR 2444076.
40. Fioretti & Alfani; Alfani, A (1988). "Anatomy of Succulence and CAM in 15 Species of Senecio". Botanical Gazette. The University of Chicago Press. 149 (2): 142–152. doi:10.1086/337701. JSTOR 2995362.
41. Holtum, Winter, Weeks and Sexton (2007). "Crassulacean acid metabolism in the ZZ plant, Zamioculcas zamiifolia (Araceae)". American Journal of Botany. 94 (10): 1670–1676. doi:10.3732/ajb.94.10.1670.
42. Crayn, D. M.; Winter, K; Smith, JA (2004). "Multiple origins of crassulacean acid metabolism and the epiphytic habit in the Neotropical family Bromeliaceae". Proceedings of the National Academy of Sciences. 101 (10): 3703–8. doi:10.1073/pnas.0400366101. PMC 373526 . PMID 14982989.

CHAPTER 8

Rubisco Process

Chapter Contents

An enzyme is a type of protein that makes reactions occur faster. For example, there is an enzyme in our body that helps transfer carbon dioxide from our cells to our blood. This reaction occurs 107 times faster with the enzyme than without.

The Calvin Cycle is a part of photosynthesis where plants take water, carbon dioxide, and light and turn them into sugar and oxygen. The Calvin Cycle starts with carbon dioxide and ends with sugar.

INTRODUCTION

Ribulose-1,5-bisphosphate carboxylase/oxygenase, commonly known by the abbreviations RuBisCO, RuBPCase, or RuBPco, is an enzyme involved in the first major step of carbon fixation, a process by which atmospheric carbon dioxide is converted by plants and other photosynthetic organisms to energy-rich molecules such as glucose. In chemical terms, it catalyzes the carboxylation of ribulose-1,5-bisphosphate. It is probably the most abundant enzyme on Earth.

STRUCTURE

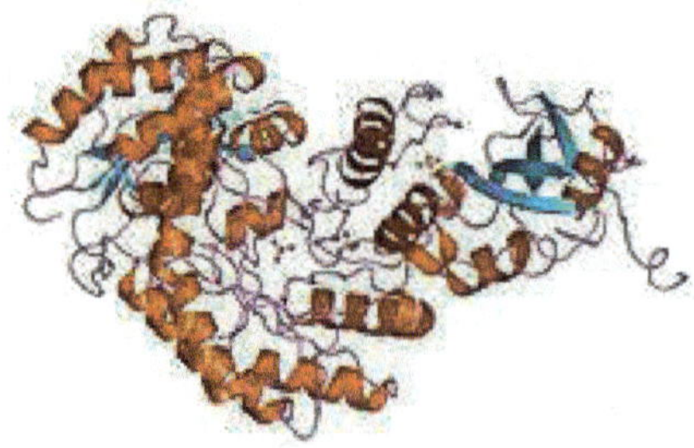

PDB structure 3rub: the ribulose-1,5-bisphosphate carboxylase/oxygenase from tobacco.

In plants, algae, cyanobacteria, and phototrophic and chemoautotrophic proteobacteria, the enzyme usually consists of two types of protein subunit, called the large chain (**L**, about 55,000 Da) and the small chain (**S**, about 13,000 Da). The *large-chain* gene (*rbcL*) is encoded by the chloroplast DNA in plants. There are typically several related *small-chain* genes in the nucleus of plant cells, and the small chains are imported to the stromal compartment of chloroplasts from the cytosol by crossing the outer chloroplast membrane. The enzymatically active substrate (ribulose 1,5-bisphosphate) binding sites are located in the large chains that form dimers as shown in Figure 1 (above, right) in which amino acids from each large chain contribute to the binding sites. A total of eight large-chains (= 4 dimers) and eight small chains assemble into a larger complex of about 540,000 Da. In some proteobacteria and dinoflagellates, enzymes consisting of only large subunits have been found.

Magnesium ions ($Mg2^{+}$) are needed for enzymatic activity. Correct positioning of $Mg2^{+}$ in the active site of the enzyme involves addition of an "activating" carbon dioxide molecule (CO_2) to a lysine in the active site (forming a carbamate). Formation of the carbamate is favored by an alkaline pH. The pH and the concentration of magnesium ions in the fluid compartment (in plants, the stroma of the chloroplast) increases in the light. The role of changing pH and magnesium ion levels in the regulation of RuBisCO enzyme activity is discussed below.

RuBisCO large chain, catalytic domain	
Identifiers	
Symbol	RuBisCO_large
Pfam	PF00016
InterPro	IPR000685
PROSITE	PDOC00142
SCOP	3rub
SUPERFAMILY	3rub

CDD	cd08148

Available protein structures:	
Pfam	structures
PDB	RCSB PDB; PDBe; PDBj
PDBsum	structure summary

RuBisCO, N-terminal domain	
Identifiers	
Symbol	RuBisCO_large_N
Pfam	PF02788
InterPro	IPR017444
SCOP	3rub
SUPERFAMILY	3rub

Available protein structures:	
Pfam	structures
PDB	RCSB PDB; PDBe; PDBj
PDBsum	structure summary

RuBisCO, small chain	
Identifiers	
Symbol	RuBisCO_small
Pfam	PF00101
InterPro	IPR000894
SCOP	3rub
SUPERFAMILY	3rub
CDD	cd03527

Available protein structures:	
Pfam	structures
PDB	RCSB PDB; PDBe; PDBj
PDBsum	structure summary

Enzymatic Activity

RuBisCO is one of many enzymes in the Calvin cycle.

Substrates

During carbon fixation, the substrate molecules for RuBisCO are ribulose-1,5-bisphosphate and carbon dioxide (distinct from the "activating" carbon dioxide). RuBisCO also catalyses a reaction between ribulose-1,5-bisphosphate and molecular oxygen (O_2) instead of carbon dioxide (CO_2).

Products

When carbon dioxide is the substrate, the product of the carboxylase reaction is a highly unstable six-carbon phosphorylated intermediate known as 3-keto-2-carboxyarabinitol-1,5-bisphosphate, which decays virtually instantaneously into two molecules of glycerate-3-phosphate. The extremely unstable molecule created by the initial carboxylation was unknown until 1988, when it was isolated. The 3-phosphoglycerate can be used to produce larger molecules such as glucose. Also, Rubisco side activities can lead to useless or inhibitory by-products; one such product is xylulose-1,5-bisphosphate, which inhibits Rubisco activity.

When molecular oxygen is the substrate, the products of the oxygenase reaction are phosphoglycolate and 3-phosphoglycerate. Phosphoglycolate is recycled through a sequence of reactions called photorespiration, which involves enzymes and cytochromes located in the mitochondria and peroxisomes (this is a case of metabolite repair). In this process, two molecules of phosphoglycolate are converted to one molecule of carbon dioxide and one molecule of 3-phosphoglycerate, which can reenter the Calvin cycle. Some of the phosphoglycolate entering this pathway can be retained by plants to produce other molecules such as glycine. At ambient levels of carbon dioxide and oxygen, the ratio of the reactions is about 4 to 1, which results in a net carbon dioxide fixation of only 3.5. Thus, the inability of the enzyme to prevent the reaction with oxygen greatly reduces the photosynthetic capacity of many plants. Some plants, many algae, and photosynthetic bacteria have overcome this limitation by devising means to increase the concentration of carbon dioxide around the enzyme, including C_4 carbon fixation, crassulacean acid metabolism, and the use of pyrenoid.

Rate of Enzymatic Activity

ome enzymes can carry out thousands of chemical reactions each second. However, RuBisCO is slow, being able to fix only 3-10 carbon dioxide molecules each second per molecule of enzyme. The reaction catalyzed by RuBisCO is, thus, the primary rate-limiting factor of the Calvin cycle during the day. Nevertheless, under most conditions, and when light is not otherwise limiting photosynthesis, the speed of RuBisCO responds positively to increasing carbon dioxide concentration. However, our descriptive knowledge will become more usable when we can translate them into quantitative models that can enable us to calculate the outcome of the reaction under a given condition. Since RubisCO reacts with RuBP (ribulose 1,5 bisphosphate) first to produces enediol and next with CO_2 that after some intermediate changes produces PGA (3-phosphoglycerate), a biochemical model is developed to represent the effects of these steps quantitatively. Since carboxylation or fixation of CO_2 is possible only after the synthesis of enediol, thus it is suggested that the role of RubisCO is to produce enediol that is carboxylase and oxygenase (EnCO). Accordingly, RubisCO is called enolase-phosphglycerase (EPGase) since it is neither carboxylase nor oxygenase.

REGULATION OF ITS ENZYMATIC ACTIVITY

RuBisCO is usually only active during the day as ribulose 1,5-bisphosphate is not regenerated in the dark. This is due to the regulation of several other enzymes in the Calvin cycle. In addition, the activity of RuBisCO is coordinated with that of the other enzymes of the Calvin cycle in several ways.

Regulation by Ions

Upon illumination of the chloroplasts, the pH of the stroma rises from 7.0 to 8.0 because of the proton (hydrogen ion, H^+) gradient created across the thylakoid membrane. At the same time, magnesium ions ($Mg2^+$) move out of the thylakoids, increasing the concentration of magnesium in the stroma of the chloroplasts. RuBisCO has a high optimal pH (can be >9.0, depending on the magnesium ion concentration) and, thus, becomes "activated" by the addition of carbon dioxide and magnesium to the active sites as described above.

Regulation by RuBisCO Activase

In plants and some algae, another enzyme, RuBisCO activase, is required to allow the rapid formation of the critical carbamate in the active site of RuBisCO.RuBisCO activase is required because the ribulose 1,5-bisphosphate (RuBP) substrate binds more strongly to the active sites lacking the carbamate and markedly slows down the "activation" process. In the light, RuBisCO activase promotes the release of the inhibitory, or — in some views — storage RuBP from the catalytic sites. Activase is also required in some plants (e.g., tobacco and many beans) because, in darkness, RuBisCO is inhibited (or protected from hydrolysis) by a competitive inhibitor synthesized by these plants, a substrate analog 2-Carboxy-D-arabitinol 1-phosphate (CA1P). CA1P binds tightly to the active site of carbamylated RuBisCO and inhibits catalytic activity. In the light, RuBisCO activase also promotes the release of CA1P from the catalytic sites. After the CA1P is released from RuBisCO, it is rapidly converted to a non-inhibitory form by a light-activated CA1P-phosphatase. Finally, once every several hundred reactions, the normal reactions with carbon dioxide or oxygen are not completed, and other inhibitory substrate analogs are formed in the active site. Once again, RuBisCO activase can promote the release of these analogs from the catalytic sites and maintain the enzyme in a catalytically active form. The properties of activase limit the photosynthetic potential of plants at high temperatures. CA1P has also been shown to keep RuBisCO in a conformation that is protected from proteolysis. At high temperatures, RuBisCO activase aggregates and can no longer activate RuBisCO. This contributes to the decreased carboxylating capacity observed during heat stress.

Regulation by ATP/ADP and Stromal Reduction/Oxidation State through the Activase

The removal of the inhibitory RuBP, CA1P, and the other inhibitory substrate analogs by activase requires the consumption of ATP. This reaction is inhibited by the presence of ADP, and, thus, activase activity depends on the ratio of these compounds in the chloroplast stroma. Furthermore, in most plants, the sensitivity of activase to the ratio of ATP/ADP is modified by the stromal reduction/oxidation (redox) state through another small regulatory protein, thioredoxin. In this manner, the activity of activase and the activation state of RuBisCO can be modulated in response to light intensity and, thus, the rate of formation of the ribulose 1,5-bisphosphate substrate.

Regulation by Phosphate

In cyanobacteria, inorganic phosphate (P_i) participates in the co-ordinated regulation of photosynthesis. P_i binds to the RuBisCO active site and to another site on the large chain where it can influence transitions between activated and less active conformations of the enzyme. Activation of bacterial RuBisCO might be particularly sensitive to P_i levels, which can act in the same way as RuBisCO activase in higher plants.

Regulation by Carbon Dioxide

Since carbon dioxide and oxygen compete at the active site of RuBisCO, carbon fixation by RuBisCO can be enhanced by increasing the carbon dioxide level in the compartment containing RuBisCO (chloroplast stroma). Several times during the evolution of plants, mechanisms have evolved for increasing the level of carbon dioxide in the stroma (see C_4 carbon fixation). The use of oxygen as a substrate appears to be a puzzling process, since it seems to throw away captured energy. However, it may be a mechanism for preventing overload during periods of high light flux. This weakness in the enzyme is the cause of photorespiration, such that healthy leaves in bright light may have zero net carbon fixation when the ratio of O_2 to CO_2 reaches a threshold at which oxygen is fixed instead of carbon. This phenomenon is primarily temperature-dependent. High temperature decreases the concentration of CO_2 dissolved in the moisture in the leaf tissues. This phenomenon is also related to water stress. Since plant leaves are evaporatively cooled, limited water causes high leaf temperatures. C_4 plants use the enzyme PEP carboxylase initially, which has a higher affinity for CO_2. The process first makes a 4-carbon intermediate compound, which is shuttled into a site of C_3photosynthesis then de-carboxylated, releasing CO_2 to boost the concentration of CO_2, hence the name C_4 plants.

Crassulacean acid metabolism (CAM) plants keep their stomata (on the underside of the leaf) closed during the day, which conserves water but prevents the light-independent reactions (a.k.a. the Calvin Cycle) from taking place, since these reactions require CO_2 to pass by gas exchange through these openings. Evaporation through the upper side of a leaf is prevented by a layer of wax.

Genetic Engineering

Since RuBisCO is often rate-limiting for photosynthesis in plants, it may be possible to improve photosynthetic efficiency by modifying RuBisCO genes in plants to increase catalytic activity and/or decrease oxygenation rates. This could improve biosequestration of CO_2 and be an important climate change strategy. Approaches under investigation include transferring RuBisCO genes from one organism in another organism, increasing the level of expression of RuBisCO subunits, expressing RuBisCO small chains from the chloroplast DNA, and altering RuBisCO genes to increase specificity for carbon dioxide or otherwise increase the rate of carbon fixation.

One avenue is to introduce RuBisCO variants with naturally high specificity values such as the ones from the red alga *Galdieria partita* into plants. This may improve the photosynthetic efficiency of crop plants, although possible negative impacts have yet to be studied. Advances in this area include the replacement of the tobacco enzyme with that of the purple photosynthetic bacterium *Rhodospirillum rubrum*. In 2014, two transplastomic tobacco lines with functional RuBisCO from the cyanobacterium *Synechococcus elongatus* PCC7942 (Se7942) were created by replacing the RuBisCO with the large and small subunit genes of the Se7942 enzyme, in combination with either the corresponding Se7942 assembly chaperone, RbcX, or an internal carboxysomal protein, CcmM35. Both mutants had increased CO_2 fixation rates when measured as carbon molecules per RuBisCO. However, the mutant plants grew more slowly than wild-type.

A recent theory explores the trade-off between the relative specificity (i.e., ability to favour CO_2 fixation over O_2 incorporation, which leads to the energy-wasteful process of photorespiration) and the rate at which product is formed. The authors conclude that RuBisCO may actually have evolved to reach a point of ‹near-perfection› in many plants (with widely varying substrate availabilities and environmental conditions), reaching a compromise between specificity and reaction rate.

Since photosynthesis is the single most effective natural regulator of carbon dioxide in the Earth›s atmosphere, a biochemical model of RuBisCO reaction is used as the core module of climate change models. Thus, a correct model of this reaction is essential to the basic understanding of the relations and interactions of environmental models. A new theory and model of the biochemical reaction of photosynthesis and the drawbacks of today›s most widely used model of photosynthesis is discussed in volume 29 of Advances in Photosynthesis and Respiration.

Phylogenetic Studies

The chloroplast gene *rbc*L, which codes for the large subunit of RuBisCO has been widely used as an appropriate locus for analysis of phylogenetics in plant taxonomy.

History of the Term

The term "RuBisCO" was coined humorously in 1979, by David Eisenberg at a seminar honouring the retirement of the early, prominent RuBisCO researcher, Sam Wildman, and also alluded to the snack food trade name "*Nabisco*" in reference to Wildman's attempts to create an edible protein supplement from tobacco leaves.

REFERENCES

1. Dhingra A, Portis AR, Daniell H (April 2004). "Enhanced translation of a chloroplast-expressed RbcS gene restores small subunit levels and photosynthesis in nuclear RbcS antisense plants". Proc. Natl. Acad. Sci. U.S.A. 101 (16): 6315–20. Bibcode:2004PNAS..101.6315D. doi:10.1073/pnas.0400981101. PMC 395966 . PMID 15067115. (Rubisco) is the most prevalent enzyme on this planet, accounting for 30–50% of total soluble protein in the chloroplast;
2. Feller U, Anders I, Mae T (2008). "Rubiscolytics: fate of Rubisco after its enzymatic function in a cell is terminated". J. Exp. Bot. 59 (7): 1615–24. doi:10.1093/jxb/erm242. PMID 17975207.
3. Curmi PM, Cascio D, Sweet RM, Eisenberg D, Schreuder H (August 1992). "Crystal structure of the unactivated form of ribulose-1,5-bisphosphate carboxylase/oxygenase from tobacco refined at 2.0-A resolution". J. Biol. Chem. 267 (24): 16980–9. PMID 1512238.
4. (Entrez GeneID:)
5. Dhingra A, Portis AR, Daniell H (April 2004). "Enhanced translation of a chloroplast-expressed RbcS gene restores small subunit levels and photosynthesis in nuclear RbcS antisense plants". Proc. Natl. Acad. Sci. U.S.A. 101 (16): 6315–20. Bibcode:2004PNAS..101.6315D. doi:10.1073/pnas.0400981101. PMC 395966 . PMID 15067115.
6. Arabidopsis thaliana has four RuBisCO small chain genes. The pattern of how large chains and small chains assemble is illustrated in Figure 3 (right).
7. Yoon M, Putterill JJ, Ross GS, Laing WA (April 2001). "Determination of the relative expression levels of rubisco small subunit genes in Arabidopsis by rapid amplification of cDNA ends". Anal. Biochem. 291 (2): 237–44. doi:10.1006/abio.2001.5042. PMID 11401297.
8. Stryer, Lubert; Berg, Jeremy Mark; Tymoczko, John L. (2002). "20. The Calvin Cycle and the Pentose Phosphate Pathway". Biochemistry (5th ed.). San Francisco: W.H. Freeman. ISBN 0-7167-3051-0. Figure 20.3. Structure of Rubisco. (Color-coded ribbon diagram)
9. The structure of RuBisCO from the photosynthetic bacterium Rhodospirillum rubrum has been determined by X-ray crystallography, see: PDB: 9RUB. A comparison of the structures of eukaryotic and bacterial RuBisCO is shown in the Protein Data Bank feature article on Rubisco.

10. Molecular Cell Biology, 4th edition, by Harvey Lodish, Arnold Berk, S. Lawrence Zipursky, Paul Matsudaira, David Baltimore and James E. Darnell. Published by W. H. Freeman & Co. (2000) New York. Online textbook. Figure 16-48 shows a structural model of the active site, including the involvement of magnesium. The Protein Data Bank feature article on RuBisCO also includes a model of magnesium at the active site.

11. The Lodish textbook describes the localization of RuBisCO to the stromal space of chloroplasts. Figure 17-7illustrates how RuBisCO small subunits move into the chloroplast stroma and assemble with the large subunits.

12. The chemical reactions catalyzed by RuBisCO are described in the online Biochemistry textbook by Stryer et al.

13. Pearce, F. Grant (2006-11-01). "Catalytic by-product formation and ligand binding by ribulose bisphosphate carboxylases from different phylogenies". The Biochemical Journal. 399 (3): 525–534. doi:10.1042/BJ20060430. ISSN 1470-8728. PMC 1615894 . PMID 16822231.

14. Ellis J.R. (2010). "Tackling unintelligent design". Nature. 463 (7278): 164–5. Bibcode:2010Natur.463..164E. doi:10.1038/463164a. PMID 20075906.

15. Farazdaghi H. (2011). "The single-process biochemical reaction of Rubisco: A unified theory and model with the effects of irradiance, CO2 and rate-limiting step on the kinetics of C3 and C4 photosynthesis from gas exchange". BioSystems. 103 (2): 265–284. doi:10.1016/j.biosystems.2010.11.004.

16. Portis AR (2003). "Rubisco activase — Rubisco's catalytic chaperone". Photosyn. Res. 75 (1): 11–27. doi:10.1023/A:1022458108678. PMID 16245090.

17. Jin SH, Jiang DA, Li XQ, Sun JW (August 2004). "Characteristics of photosynthesis in rice plants transformed with an antisense Rubisco activase gene". J. Zhejiang Univ. Sci. 5 (8): 897–9. doi:10.1631/jzus.2004.0897. PMID 15236471.

18. Andralojc PJ, Dawson GW, Parry MA, Keys AJ (December 1994). "Incorporation of carbon from photosynthetic products into 2-carboxyarabinitol-1-phosphate and 2-carboxyarabinitol". Biochem. J. 304 (3): 781–6. PMC 1137402 . PMID 7818481.

19. Crafts-Brandner SJ, Salvucci ME (November 2000). "Rubisco activase constrains the photosynthetic potential of leaves at high temperature and CO2". Proc. Natl. Acad. Sci. U.S.A. 97 (24): 13430–5. Bibcode:2000PNAS...9713430C. doi:10.1073/pnas.230451497. PMC 27241 . PMID 11069297.

20. Khan S, Andralojc PJ, Lea PJ, Parry MA (December 1999). "2'-carboxy-D-arabitinol 1-phosphate protects ribulose 1, 5-bisphosphate carboxylase/oxygenase against proteolytic breakdown". Eur. J. Biochem. 266 (3): 840–7. doi:10.1046/j.1432-1327.1999.00913.x. PMID 10583377.

21. Salvucci ME, Osteryoung KW, Crafts-Brandner SJ, Vierling E (November 2001). "Exceptional sensitivity of Rubisco activase to thermal denaturation in vitro and in vivo". Plant Physiol. 127 (3): 1053–64. doi:10.1104/pp.010357. PMC 129275 . PMID 11706186.

22. Zhang N, Kallis RP, Ewy RG, Portis AR (March 2002). "Light modulation of Rubisco in Arabidopsis requires a capacity for redox regulation of the larger Rubisco activase isoform". Proc. Natl. Acad. Sci. U.S.A. 99 (5): 3330–4. Bibcode:2002PNAS...99.3330Z. doi:10.1073/pnas.042529999. PMC 122518 . PMID 11854454.

23. Marcus Y, Gurevitz M (October 2000). "Activation of cyanobacterial RuBP-carboxylase/oxygenase is facilitated by inorganic phosphate via two independent mechanisms". Eur. J. Biochem. 267 (19): 5995–6003. doi:10.1046/j.1432-1327.2000.01674.x. PMID 10998060.

24. Spreitzer RJ, Salvucci ME (2002). "Rubisco: structure, regulatory interactions, and possibilities for a better enzyme". Annu Rev Plant Biol. 53: 449–75. doi:10.1146/annurev.arplant.53.100301.135233. PMID 12221984.

25. Parry MA, Andralojc PJ, Mitchell RA, Madgwick PJ, Keys AJ (May 2003). "Manipulation of Rubisco: the amount, activity, function and regulation". J. Exp. Bot. 54 (386): 1321–33. doi:10.1093/jxb/erg141. PMID 12709478.

26. Whitney SM, Andrews TJ (December 2001). "Plastome-encoded bacterial ribulose-1,5-bisphosphate carboxylase/oxygenase (RubisCO) supports photosynthesis and growth in tobacco". Proc. Natl. Acad. Sci. U.S.A. 98 (25): 14738–43. Bibcode:2001PNAS...9814738W. doi:10.1073/pnas.261417298. PMC 64751 . PMID 11724961.

27. John Andrews T, Whitney SM (June 2003). "Manipulating ribulose bisphosphate carboxylase/oxygenase in the chloroplasts of higher plants". Arch. Biochem. Biophys. 414 (2): 159–69. doi:10.1016/S0003-9861(03)00100-0. PMID 12781767.

28. Lin, M. T.; Occhialini, A.; Andralojc, P. J.; Parry, M. A. J.; Hanson, M. R. (2014). "A faster RuBisCO with potential to increase photosynthesis in crops". Nature. 513 (7519): 547. doi:10.1038/nature13776. PMID 25231869.

29. Tcherkez GG, Farquhar GD, Andrews TJ (May 2006). "Despite slow catalysis and confused substrate specificity, all ribulose bisphosphate carboxylases may be nearly perfectly optimized". Proc. Natl. Acad. Sci. U.S.A. 103 (19): 7246–51. Bibcode:2006PNAS..103.7246T. doi:10.1073/pnas.0600605103. PMC 1464328 . PMID 16641091. Archived from the original on April 6, 2015.

30. Farazdaghi, Hadi (2009). "Modeling the Kinetics of Activation and Reaction of Rubisco from Gas Exchange (12)". In Laisk A; Nedbal L; Govindjee. Photosynthesis in silico: Understanding Complexity from Molecules to Ecosystems. Advances in Photosynthesis and Respiration. 29. Berlin: Springer. ISBN 1-4020-9236-9.

31. Chase et al 1993.

32. Wildman SG (2002). "Along the trail from Fraction I protein to Rubisco (ribulose bisphosphate carboxylase-oxygenase)". Photosyn. Res. 73 (1-3): 243–50. doi:10.1023/A:1020467601966. PMID 16245127.

33. Portis AR, Parry MA (October 2007). "Discoveries in Rubisco (Ribulose 1,5-bisphosphate carboxylase/oxygenase): a historical perspective". Photosyn. Res. 94 (1): 121–43. doi:10.1007/s11120-007-9225-6. PMID 17665149.

CHAPTER 9

Emerson & Warburg Effect

Chapter Contents

- **Introduction**
- **Description**
- **Warburg Effect**

Emerson Effect: (enhancement effect) The observation (made by Robert Emerson in 1957) that photosynthesis, which proceeds very slowly using light of 700 nm wavelength, can be greatly increased when chloroplasts are also illuminated with light of shorter wavelength (650 nm). The EMERSON - Effect: The rate of photosynthesis after exposition to light of the wave length lambda = 670 nm and lambda = 700 nm, respectively. When simultaneously exposed to light of both wave lengths is the rate of photosynthesis increased (EMERSON - effect, at the right).

INTRODUCTION

The Emerson effect is the increase in the rate of photosynthesis after chloroplasts are exposed to light of wavelength 670 nm (red light) and 700 nm (far red light). When simultaneously exposed to light of both wavelengths, the rate of photosynthesis is far higher than the sum of the red light and far red light photosynthesis rates. The effect was early evidence that two photosystems, processing different wavelengths, cooperate in photosynthesis.

Robert Emerson described the eponymous effect in 1957. In his paper he observed that:

- When plants are exposed to light having wavelength greater than 680 nm, then only one photosystem is activated; i.e. PS700 resulting in formation of ATP only.
- When plants are exposed to light having wavelength less than 680 nm, the rate of photosynthesis was very low.
- On giving both shorter and higher wavelengths of light, the efficiency of the process increased, because both photosystems were working together at the same time, resulting in higher yield.

DESCRIPTION

When Emerson exposed green plants to differing wavelengths of light, he noticed that at wavelengths of greater than 680 nm the efficiency of photosynthesis decreased abruptly despite the fact that this is a region of the spectrum where chlorophyll still absorbs light (chlorophyll is the green pigment in plants - it absorbs mainly the red and blue wavelengths from light, leaving the green light to bounce back and hit our eyes). When the plants were exposed to short-wavelength light, (less than 660 nm), the efficiency also decreased. Emerson then exposed the plants to both short and long wavelengths at the same time, causing the efficiency to increase greatly. He concluded that there must be two different photosystems involved in photosynthesis, one driven by short-wavelength light and one driven by long-wavelength (PS1 and PS2). They work together to enhance efficiency and convert the light energy to forms that can be absorbed by the plant.

The light excites the chlorophyll molecules at the reaction centre and causes an increase in energy. As the molecule becomes less excited, its energy is transported through a chain of electron carriers to the next photosystem which does much the same thing and produces energy-carrying organic molecules.

WARBURG EFFECT

The phrase "Warburg effect" is used for two unrelated observations in biochemistry, one in plant physiology and the other in oncology, both due to Nobel laureate Otto Heinrich Warburg.

Plant Physiology

In plant physiology, the Warburg's effect is the decrease in the rate of photosynthesis by high oxygen concentrations. Oxygen is a competitive inhibitor of the carbon dioxide fixation by RuBisCO which initiates photosynthesis. Furthermore, oxygen stimulates photorespiration which reduces photosynthetic output. These two mechanisms working together are responsible for the Warburg effect.

Oncology

Basis

In oncology, the Warburg effect is the observation that most cancer cells predominantly produce energy by a high rate of glycolysis followed by lactic acid fermentation in the cytosol, rather than by a comparatively low rate of glycolysis followed by oxidation of pyruvate in mitochondria as in most normal cells.The latter process is aerobic (uses oxygen). Malignant, rapidly growing tumor cells typically have glycolytic rates up to 200 times higher than those of their normal tissues of origin; this occurs even if oxygen is plentiful.

Otto Warburg postulated this change in metabolism is the fundamental cause of cancer, a claim now known as the Warburg hypothesis. Today, mutations in oncogenes and tumor suppressor genes are thought to be responsible for malignant transformation, and the Warburg effect is considered to be a result of these mutations rather than a cause.

Possible Explanations

The Warburg effect may simply be a consequence of damage to the mitochondria in cancer, or an adaptation to low-oxygen environments within tumors, or a result of cancer genes shutting down the mitochondria because they are involved in the cell's apoptosis program which would otherwise kill cancerous cells. It may also be an effect associated with cell proliferation. Since glycolysis provides most of the building blocks required for cell proliferation, cancer cells (and normal proliferating cells) have been proposed to need to activate glycolysis, despite the presence of oxygen, to proliferate. Evidence attributes some of the high aerobic glycolytic rates to an overexpressed form of mitochondrially-bound hexokinase responsible for driving the high glycolytic activity. In kidney cancer, this effect could be due to the presence of mutations in the von Hippel–Lindau tumor suppressor gene upregulating glycolytic enzymes, including the M2 splice isoform of pyruvate kinase.

In March 2008, Lewis C. Cantley and colleagues announced that the tumor M2-PK, a form of the pyruvate kinase enzyme, gives rise to the Warburg effect. Tumor M2-PK is produced in all rapidly dividing cells, and is responsible for enabling cancer cells to consume glucose at an accelerated rate; on forcing the cells to switch to pyruvate kinase's alternative form by inhibiting the production of tumor M2-PK, their growth was curbed. The researchers acknowledged the fact that the exact chemistry of glucose metabolism was likely to vary across different forms of cancer; but PKM2 was identified in all of the cancer cells they had tested. This enzyme form is not usually found in healthy tissue, though it is apparently necessary when cells need to multiply quickly, e.g. in healing wounds or hematopoiesis.

Glycolytic Inhibitors

Many substances have been developed which inhibit glycolysis, and such inhibitors are currently the subject of intense research as anticancer agents, including SB-204990, 2-deoxy-D-glucose (2DG), 3-bromopyruvate (3-BrPA, bromopyruvic acid, or bromopyruvate), 3-BrOP, 5-thioglucose and dichloroacetic acid (DCA). Clinical trials are ongoing for 2-DG and DCA.

Alpha-cyano-4-hydroxycinnamic acid (ACCA;CHC), a small-molecule inhibitor of monocarboxylate transporters (MCTs; which prevent lactic acid build up in tumors) has been successfully used as a metabolic target in brain tumor pre-clinical research. Higher affinity MCT inhibitors have been developed and are currently undergoing clinical trials by Astra-Zeneca.

Dichloroacetic acid (DCA), a small-molecule inhibitor of mitochondrial pyruvate dehydrogenase kinase, "downregulates" glycolysis *in vitro* and *in vivo*. Researchers at the University of Alberta theorized in 2007 that DCA might have therapeutic benefits against many types of cancers.

Blood Glucose Levels

High glucose levels have been shown to accelerate cancer cell proliferation in vitro, while glucose deprivation has led to apoptosis. These findings have initiated further study of the effects of carbohydrate restriction on tumor growth. Clinical evidence shows that lower blood glucose levels in late-stage cancer patients have been correlated with better outcomes.

Alternative Models

A model called the reverse Warburg effect describes cells producing energy by glycolysis, but were not tumor cells, but stromal fibroblasts. Although the Warburg effect would exist in certain cancer types potentially, it highlighted the need for a closer look at tumor metabolism.

Metabolic reprogramming is also observed in neurodegenerative diseases, Alzheimer's and Parkinson's. This metabolic alteration is described by the up-regulation of oxidative phosphorylation - called the inverse Warburg Effect.

Cancer Metabolism and Epigenetics

Nutrient utilization is dramatically altered when cells receive signals to proliferate. Characteristic metabolic changes enable cells to meet the large biosynthetic demands associated with cell growth and division. Changes in rate-limiting glycolytic enzymes redirect metabolism to support growth and proliferation. Metabolic reprogramming in cancer is largely due to oncogenic activation of signal transduction pathways and transcription factors. Although less well understood, epigenetic mechanisms also contribute to the regulation of metabolic gene expression in cancer. Reciprocally, accumulating evidence suggest that metabolic alterations may affect the epigenome. Understanding the relation between metabolism and epigenetics in cancer cells may open new avenues for anti-cancer strategies.

REFERENCES

1. Govindjee; Rabinowitch, Eugene (1969). Photosynthesis (1 ed.). New York City: John Wiley & Sons, Inc. p. 172. ISBN 0-471-70423-7.
2. Emerson, Robert (1957). "Dependence of yield of photosynthesis in long wave red on wavelength and intensity of supplementary light". Science. 125: 746.
3. plantphysiol.org
4. Turner JS, Brittain EG (February 1962). "Oxygen as a factor in photosynthesis" (PDF). Biological Reviews of the Cambridge Philosophical Society. 37: 130–70. doi:10.1111/j.1469-185X.1962.tb01607.x. PMID 13923215.
5. Zelitch I (1971). "Chapter 8, Section E: Inhibition by O2 (The Warburg Effect)". Photosynthesis, Photorespiration, and Plant Productivity. New York: Academic Press. pp. 253–255. ISBN 0124316085.
6. Schopfer P, Mohr H (1995). "The leaf as a photosynthetic system". Plant physiology. Berlin: Springer. pp. 236–237. ISBN 3-540-58016-6.
7. Alfarouk KO, Verduzco D, Rauch C, Muddathir AK, Adil HH, Elhassan GO, Ibrahim ME, David Polo Orozco J, Cardone RA, Reshkin SJ, Harguindey S (2014). "Glycolysis, tumor metabolism, cancer growth and dissemination. A new pH-based etiopathogenic perspective and therapeutic approach to an old cancer question". Oncoscience. 1 (12): 777–802. doi:10.18632/oncoscience.109. PMC 4303887 . PMID 25621294.
8. Alfarouk KO (February 2016). "Tumor metabolism, cancer cell transporters, and microenvironmental resistance". Journal of Enzyme Inhibition and Medicinal Chemistry: 1–8. doi:10.3109/14756366.2016.1140753. PMID 26864256.
9. Alfarouk KO, Muddathir AK, Shayoub ME (20 January 2011). "Tumor acidity as evolutionary spite". Cancers. 3 (1): 408–14. doi:10.3390/cancers3010408. PMC 3756368 . PMID 24310355.
10. Gatenby RA, Gillies RJ (November 2004). "Why do cancers have high aerobic glycolysis?". Nature Reviews. Cancer. 4 (11): 891–9. doi:10.1038/nrc1478. PMID 15516961.
11. Kim JW, Dang CV (September 2006). "Cancer's molecular sweet tooth and the Warburg effect". Cancer Research. 66 (18): 8927–30. doi:10.1158/0008-5472.CAN-06-1501. PMID 16982728.
12. Warburg O (February 1956). "On the origin of cancer cells". Science. 123 (3191): 309–14. Bibcode:1956Sci...123..309W. doi:10.1126/science.123.3191.309. PMID 13298683.
13. Bertram JS (December 2000). "The molecular biology of cancer". Molecular Aspects of Medicine. 21 (6): 167–223. doi:10.1016/S0098-2997(00)00007-8. PMID 11173079.

14. Grandér D (April 1998). "How do mutated oncogenes and tumor suppressor genes cause cancer?". Medical Oncology. 15 (1): 20–6. doi:10.1007/BF02787340. PMID 9643526.
15. López-Lázaro M (April 2008). "The warburg effect: why and how do cancer cells activate glycolysis in the presence of oxygen?". Anti-Cancer Agents in Medicinal Chemistry. 8 (3): 305–12. doi:10.2174/187152008783961932. PMID 18393789.
16. Bustamante E, Pedersen PL (September 1977). "High aerobic glycolysis of rat hepatoma cells in culture: role of mitochondrial hexokinase". Proceedings of the National Academy of Sciences of the United States of America. 74 (9): 3735–9. Bibcode:1977PNAS...74.3735B. doi:10.1073/pnas.74.9.3735. PMC 431708 . PMID 198801.
17. Unwin RD, Craven RA, Harnden P, Hanrahan S, Totty N, Knowles M, Eardley I, Selby PJ, Banks RE (August 2003). "Proteomic changes in renal cancer and co-ordinate demonstration of both the glycolytic and mitochondrial aspects of the Warburg effect". Proteomics. 3 (8): 1620–32. doi:10.1002/pmic.200300464. PMID 12923786.
18. Christofk HR, Vander Heiden MG, Harris MH, Ramanathan A, Gerszten RE, Wei R, Fleming MD, Schreiber SL, Cantley LC (March 2008). "The M2 splice isoform of pyruvate kinase is important for cancer metabolism and tumour growth". Nature. 452 (7184): 230–3. Bibcode:2008Natur.452..230C. doi:10.1038/nature06734. PMID 18337823.
19. Pedersen PL (June 2007). "Warburg, me and Hexokinase 2: Multiple discoveries of key molecular events underlying one of cancers' most common phenotypes, the "Warburg Effect", i.e., elevated glycolysis in the presence of oxygen". Journal of Bioenergetics and Biomembranes. 39 (3): 211–22. doi:10.1007/s10863-007-9094-x. PMID 17879147.
20. Pelicano H, Martin DS, Xu RH, Huang P (August 2006). "Glycolysis inhibition for anticancer treatment". Oncogene. 25 (34): 4633–46. doi:10.1038/sj.onc.1209597. PMID 16892078.
21. Clinical trial number NCT00633087 for "A Phase I/II Trial of 2-Deoxyglucose (2DG) for the Treatment of Advanced Cancer and Hormone Refractory Prostate Cancer (2-Deoxyglucose)" at ClinicalTrials.gov
22. Colen CB (2005). Gene therapy and radiation of malignant glioma by targeting glioma specific lactate transporter (Ph.D.). Wayne State University.
23. Colen CB, Seraji-Bozorgzad N, Marples B, Galloway MP, Sloan AE, Mathupala SP (December 2006). "Metabolic remodeling of malignant gliomas for enhanced sensitization during radiotherapy: an in vitro study". Neurosurgery. 59 (6): 1313–23; discussion 1323–4. doi:10.1227/01.NEU.0000249218.65332.BF. PMC 3385862 . PMID 17277695.
24. Colen CB, Shen Y, Ghoddoussi F, Yu P, Francis TB, Koch BJ, Monterey MD, Galloway MP, Sloan AE, Mathupala SP (July 2011). "Metabolic targeting of

lactate efflux by malignant glioma inhibits invasiveness and induces necrosis: an in vivo study". Neoplasia. 13 (7): 620–32. doi:10.1593/neo.11134. PMC 3132848 . PMID 21750656.

25. Mathupala SP, Colen CB, Parajuli P, Sloan AE (February 2007). "Lactate and malignant tumors: a therapeutic target at the end stage of glycolysis". Journal of Bioenergetics and Biomembranes. 39 (1): 73–7. doi:10.1007/s10863-006-9062-x. PMC 3385854 . PMID 17354062.
26. Clinical trial number NCT01791595 for "A Phase I Trial of AZD3965 in Patients With Advanced Cancer" at ClinicalTrials.gov
27. Bonnet S, Archer SL, Allalunis-Turner J, Haromy A, Beaulieu C, Thompson R, Lee CT, Lopaschuk GD, Puttagunta L, Bonnet S, Harry G, Hashimoto K, Porter CJ, Andrade MA, Thebaud B, Michelakis ED (January 2007). "A mitochondria-K+ channel axis is suppressed in cancer and its normalization promotes apoptosis and inhibits cancer growth". Cancer Cell. 11 (1): 37–51. doi:10.1016/j.ccr.2006.10.020. PMID 17222789.
28. Pan JG, Mak TW (April 2007). "Metabolic targeting as an anticancer strategy: dawn of a new era?". Science's STKE. 2007 (381): pe14. doi:10.1126/stke.3812007pe14. PMID 17426345.
29. Klement RJ, Kämmerer U (2011). "Is there a role for carbohydrate restriction in the treatment and prevention of cancer?". Nutrition & Metabolism. 8: 75. doi:10.1186/1743-7075-8-75. PMID 22029671.
30. Pavlides S, Whitaker-Menezes D, Castello-Cros R, Flomenberg N, Witkiewicz AK, Frank PG, Casimiro MC, Wang C, Fortina P, Addya S, Pestell RG, Martinez-Outschoorn UE, Sotgia F, Lisanti MP (December 2009). "The reverse Warburg effect: aerobic glycolysis in cancer associated fibroblasts and the tumor stroma". Cell Cycle. 8 (23): 3984–4001. doi:10.4161/cc.8.23.10238. PMID 19923890.
31. Alfarouk KO, Shayoub ME, Muddathir AK, Elhassan GO, Bashir AH (22 July 2011). "Evolution of Tumor Metabolism might Reflect Carcinogenesis as a Reverse Evolution process (Dismantling of Multicellularity)". Cancers. 3 (4): 3002–17. doi:10.3390/cancers3033002. PMC 3759183 . PMID 24310356.
32. Gupta V, Gopinath P, Iqbal MA, Mazurek S, Wellen KE, Bamezai RN (2013). "Interplay between epigenetics & cancer metabolism". Current Pharmaceutical Design. 20 (11): 1706–14. doi:10.2174/13816128113199990536. PMID 23888952.

CHAPTER 10

Photosynthetic Pigments

Chapter Contents

Pigments are chemical compounds which reflect only certain wavelengths of visible light. This makes them appear "colorful". Flowers, corals, and even animal skin contain pigments which give them their colors. More important than their reflection of light is the ability of pigments to absorb certain wavelengths.

INTRODUCTION

A photosynthetic pigment (accessory pigment; chloroplast pigment; antenna pigment) is a pigment that is present in chloroplasts or photosynthetic bacteria and captures the light energy necessary for photosynthesis.

PLANTS

Green plants have six closely related photosynthetic pigments (in order of increasing polarity):

- Carotene - an orange pigment
- Xanthophyll - a yellow pigment

- Phaeophytin *a* - a gray-brown pigment
- Phaeophytin *b* - a yellow-brown pigment
- Chlorophyll *a* - a blue-green pigment
- Chlorophyll *b* - a yellow-green pigment

Chlorophyll a is the most common of the six, present in every plant that performs photosynthesis. The reason that there are so many pigments is that each absorbs light more efficiently in a different part of the electromagnetic spectrum. Chlorophyll a absorbs well at a wavelength of about 400-450 nm and at 650-700 nm; chlorophyll b at 450-500 nm and at 600-650 nm. Xanthophyll absorbs well at 400-530 nm. However, none of the pigments absorbs well in the green-yellow region, which is responsible for the abundant green we see in nature.

Carotene

The term carotene (also carotin, from the Latin *carota*, "carrot") is used for many related unsaturated hydrocarbon substances having the formula $C_{40}H_x$, which are synthesized by plants but in general cannot be made by animals (with the sole known exception of some aphids and spider mites which acquired the synthetic genes from fungi). Carotenes are photosynthetic pigments important for photosynthesis. Carotenes contain no oxygen atoms. They absorb ultraviolet, violet, and blue light and scatter orange or red light, and (in low concentrations) yellow light.

Carotenes are responsible for the orange colour of the carrot, for which this class of chemicals is named, and for the colours of many other fruits, vegetables and fungi (for example, sweet potatoes, chanterelle and orange cantaloupe melon). Carotenes are also responsible for the orange (but not all of the yellow) colours in dry foliage. They also (in lower concentrations) impart the yellow coloration to milk-fat and butter. Omnivorous animal species which are relatively poor converters of coloured dietary carotenoids to colourless retinoids have yellowed-coloured body fat, as a result of the carotenoid retention from the vegetable portion of their diet. The typical yellow-coloured fat of humans and chickens is a result of fat storage of carotenes from their diets.

Carotenes contribute to photosynthesis by transmitting the light energy they absorb to chlorophyll. They also protect plant tissues by helping to absorb the energy from singlet oxygen, an excited form of the oxygen molecule O_2 which is formed during photosynthesis.

β-Carotene is composed of two retinyl groups, and is broken down in the mucosa of the human small intestine by β-carotene 15,15'-monooxygenase to retinal, a form of vitamin A. β-Carotene can be stored in the liver and body fat and converted to retinal as needed, thus making it a form of vitamin A for humans and some other mammals. The carotenes α-carotene and γ-carotene, due to their single retinyl group (β-ionone ring), also have some vitamin A activity (though less than β-carotene), as does the xanthophyll carotenoid β-cryptoxanthin. All other carotenoids, including lycopene, have no beta-ring and thus **no** vitamin A activity (although they may have antioxidant activity and thus biological activity in other ways).

Animal species differ greatly in their ability to convert retinyl (beta-ionone) containing carotenoids to retinals. Carnivores in general are poor converters of dietary ionone-containing carotenoids. Pure carnivores such as ferrets lack β-carotene 15,15›-monooxygenase and cannot convert any carotenoids to retinals at all (resulting in carotenes **not** being a form of vitamin A for this species); while cats can convert a trace of β-carotene to retinol, although the amount is totally insufficient for meeting their daily retinol needs.

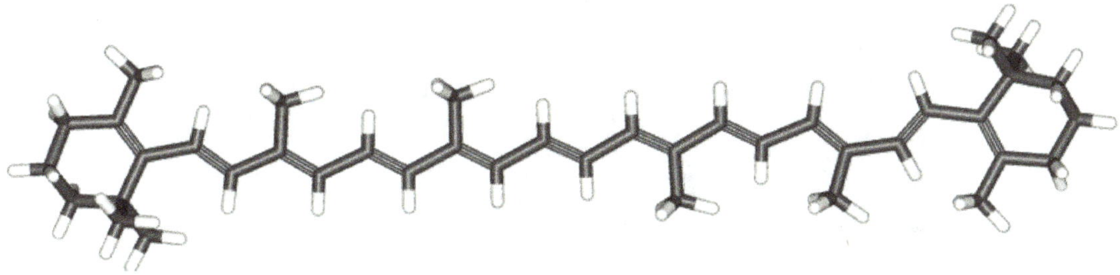

β-Carotene represented by a 3-dimensional stick diagram

Carotene is responsible for the orange colour of carrots and the colours of many other fruits and vegetables and even some animals.

Lesser Flamingos in the Ngorongoro Crater, Tanzania. The pink colour of wild flamingos is due to astaxanthin (a carotenoid) they absorb from their diet of brine shrimp. If fed a carotene-free diet they become white.

Molecular Structure

Chemically, carotenes are polyunsaturated hydrocarbons containing 40 carbon atoms per molecule, variable numbers of hydrogen atoms, and no other elements. Some carotenes are terminated by hydrocarbon rings, on one or both ends of the molecule. All are coloured to the human eye, due to extensive systems of conjugated double bonds. Structurally carotenes are tetraterpenes, meaning that they are synthesized biochemically from four 10-carbon terpene units, which in turn are formed from eight 5-carbon isoprene units.

Carotenes are found in plants in two primary forms designated by characters from the Greek alphabet: alpha-carotene (α-carotene) and beta-carotene (β-carotene). Gamma-, delta-, epsilon-, and zeta-carotene (γ, δ, ε, and ζ-carotene) also exist. Since they are hydrocarbons, and therefore contain no oxygen, carotenes are fat-soluble and insoluble in water (in contrast with other carotenoids, the xanthophylls, which contain oxygen and thus are less chemically hydrophobic).

Dietary Sources

The following foods are particularly rich in carotenes (also see Vitamin A article for amounts):

- sweet potatoes
- carrots
- wolfberries (goji)
- cantaloupe melon

- mangoes
- apricots
- Persimmon
- spinach
- kale
- chard
- turnip greens
- dandelion greens
- beet greens
- mustard greens
- collard greens
- watercress
- cilantro (coriander)
- fresh thyme
- broccoli
- parsley
- romaine lettuce
- ivy gourd
- rose hips
- winter squash
- pumpkin
- cassava

Absorption from these foods is enhanced if eaten with fats, as carotenes are fat soluble, and if the food is cooked for a few minutes until the plant cell wall splits and the colour is released into any liquid. 6 μg of dietary β-carotene supplies the equivalent of 1 μg of retinol, or 1 RE (Retinol Equivalent). This is equivalent to 3⅓ IU of vitamin A.

The Multiple Forms

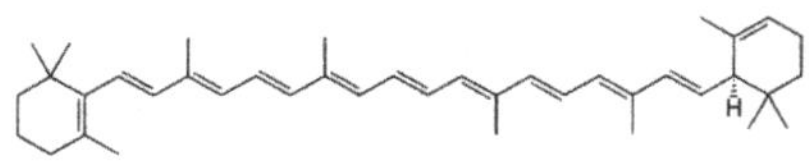

α-carotene

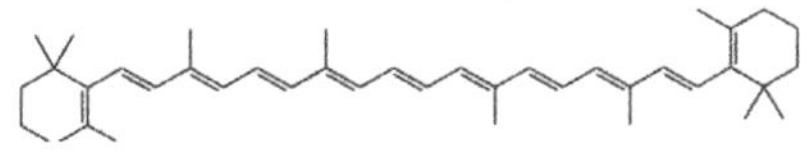

β-carotene

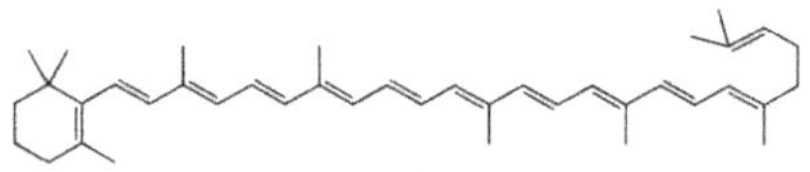

γ-carotene

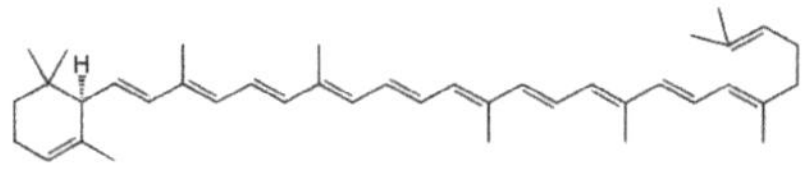

δ-carotene

The two primary isomers of carotene, α-carotene and β-carotene, differ in the position of a double bond (and thus a hydrogen) in the cyclic group at one end (the right end in the diagram at right).

β-Carotene is the more common form and can be found in yellow, orange, and green leafy fruits and vegetables. As a rule of thumb, the greater the intensity of the orange colour of the fruit or vegetable, the more β-carotene it contains.

Carotene protects plant cells against the destructive effects of ultraviolet light. β-Carotene is an antioxidant.

β-Carotene and Cancer

It has been shown in trials that the ingestion of β-carotene supplements at about 30 mg/day (10 times the Reference Daily Intake) increases the rate of lung and prostate cancer development in smokers and people with a history of asbestos exposure.

An article on the American Cancer Society says that The Cancer Research Campaign has called for warning labels on β-carotene supplements to caution smokers that such supplements may increase the risk of lung cancer.

The New England Journal of Medicine published an article in 1994 about a trial which examined the relationship between daily supplementation of β-carotene and vitamin E (α-tocopherol) and the incidence of lung cancer. The study was done using supplements and researchers were aware of the epidemiological correlation between carotenoid-rich fruits and vegetables and lower lung cancer rates. The research concluded that no reduction in lung cancer was found in the participants using these supplements, and furthermore, these supplements may, in fact, have harmful effects.

The Journal of the National Cancer Institute and The New England Journal of Medicine published articles in 1996 about a trial that was conducted to determine if vitamin A (in the form of retinyl palmitate) and β-carotene had any beneficial effects to prevent cancer. The results indicated an increased risk of lung cancer for the participants who consumed the β-carotene supplement and who had lung irritation from smoking or asbestos exposure, causing the trial to be stopped early.

A review of all randomized controlled trials in the scientific literature by the Cochrane Collaboration published in *JAMA* in 2007 found that synthetic β-carotene *increased* mortality by something between 1 and 8% (Relative Risk 1.05, 95% confidence interval 1.01–1.08). However, this meta-analysis included two large studies of smokers, so it is not clear that the results apply to the general population. The review only studied the influence of synthetic antioxidants and the results should not be translated to potential effects of fruits and vegetables.

β-Carotene and Cognition

A recent report demonstrated that 50 mg of β-carotene every other day prevented cognitive decline in a study of over 4000 physicians at a mean treatment duration of 18 years.

β-Carotene and Photosensitivity

Oral β-carotene is prescribed to people suffering from erythropoietic protoporphyria. It provides them some relief from photosensitivity.

β-Carotene and Nanotechnology

β-Carotene and lycopene molecules can be encapsulated into carbon nanotubes enhancing the optical properties of carbon nanotubes. Efficient energy transfer occurs between the encapsulated dye and nanotube — light is absorbed by the dye and without significant loss is transferred to the single wall carbon nanotube (SWCNT). Encapsulation increases chemical and thermal stability of carotene molecules; it also allows their isolation and individual characterization.

Carotenemia

Carotenemia or hypercarotenemia is excess carotene, but unlike excess vitamin A, carotene is non-toxic. Although hypercarotenemia is not particularly dangerous, it can lead to an oranging of the skin (carotenodermia), but not the conjunctiva of eyes (thus easily distinguishing it visually from jaundice). It is most commonly associated with consumption of an abundance of carrots, but it also can be a medical sign of more dangerous conditions.

Production

Algae farm ponds in Whyalla, South Australia, used to produce β-carotene.

Most of the world's synthetic supply of carotene comes from a manufacturing complex located in Freeport, Texas and owned by DSM. The other major supplier BASF also uses a chemical process to produce β-carotene. Together these suppliers account for about 85% of the β-carotene on the market. In Spain Vitatene produces natural β-carotene from fungus Blakeslea trispora, as does DSM but at much lower amount when compared to its synthetic β-carotene operation. In Australia, organic β-carotene is produced by Aquacarotene Limited from dried marine algae *Dunaliella salina* grown in harvesting ponds situated in Karratha, Western Australia. BASF Australia is also producing β-carotene from microalgae grown in two sites in Australia that are the world's largest algae farms. In Portugal, the industrial biotechnology company Biotrend is producing natural all-*trans*-β-carotene from a non genetically modified bacteria of the *Sphingomonas* genus isolated from soil.

Carotenes are also found in palm oil, corn, and in the milk of dairy cows, causing cow›s milk to be light yellow, depending on the feed of the cattle, and the amount of fat in the milk (high-fat milks, such as those produced by Guernsey cows, tend to be yellower because their fat content causes them to contain more carotene).

Carotenes are also found in some species of termites, where they apparently have been picked up from the diet of the insects.

Total Synthesis

There are currently two commonly used methods of total synthesis of β-carotene. The first was developed by the Badische Anilin- & Soda-Fabrik (BASF) and is based on the Wittig reaction with Wittig himself as patent holder:

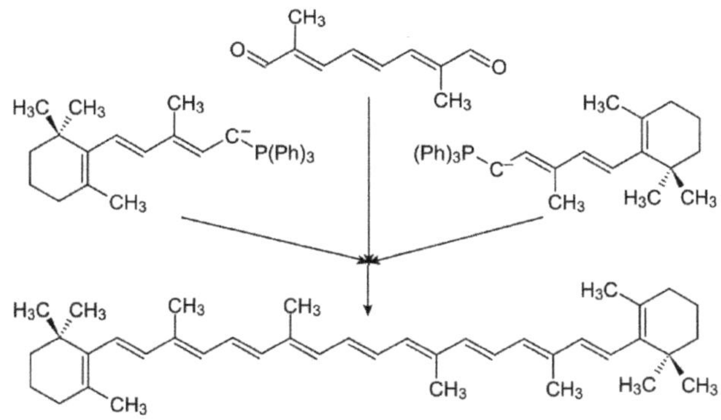

The second is a Grignard reaction, elaborated by Hoffman-La Roche from the original synthesis of Inhoffen et al. They are both symmetrical; the BASF synthesis is C20 + C20, and the Hoffman-La Roche synthesis is C19 + C2 + C19.

Nomenclature

Carotenes are carotenoids containing no oxygen. Carotenoids containing some oxygen are known as xanthophylls.

The two ends of the β-carotene molecule are structurally identical, and are called **β-rings**. Specifically, the group of nine carbon atoms at each end form a β-ring.

The α-carotene molecule has a β-ring at one end; the other end is called an **ε-ring**. There is no such thing as an "α-ring".

These and similar names for the ends of the carotenoid molecules form the basis of a systematic naming scheme, according to which:

- α-carotene is **β,ε-carotene**;
- β-carotene is **β,β-carotene**;
- γ-carotene (with one β ring and one uncyclized end that is labelled *psi*) is **β,ψ-carotene**;
- δ-carotene (with one ε ring and one uncyclized end) is **ε,ψ-carotene**;
- ε-carotene is **ε,ε-carotene**
- lycopene is **ψ,ψ-carotene**

ζ-Carotene is the biosynthetic precursor of neurosporene, which is the precursor of lycopene, which, in turn, is the precursor of the carotenes α through ε.

Food Additive

Carotene is also used as a substance to colour products such as juice, cakes, desserts, butter and margarine. It is approved for use as a food additive in the EU (listed as additive E160a) Australia and New Zealand (listed as 160a) and the US.

Xanthophyll

Xanthophylls (originally phylloxanthins) are yellow pigments that occur widely in nature and form one of two major divisions of the carotenoid group; the other division is formed by the carotenes. The name is from Greek xanthos (ξανθός, «yellow») and phyllon (φύλλον, «leaf»), due to their formation of the yellow band seen in early chromatography of leaf pigments.

The color of an egg yolk is from the xanthophyll carotenoids lutein and zeaxanthin

Molecular Structure

The molecular structure of xanthophylls is similar to that of carotenes, but xanthophylls contain oxygen atoms, while *carotenes* are purely hydrocarbons with no oxygen. Xanthophylls contain their oxygen either as hydroxyl groups and/or as pairs of hydrogen atoms that are substituted by oxygen atoms acting as a bridge (epoxide). For this reason, they are more polar than the purely hydrocarbon carotenes, and it is this difference that allows their separations from carotenes in many types of chromatography. Typically, carotenes are more orange in color than xanthophylls.

Occurrence

Like other carotenoids, xanthophylls are found in highest quantity in the leaves of most green plants, where they act to modulate light energy and perhaps serve as a non-photochemical quenching agent to deal with triplet chlorophyll (an excited form of chlorophyll), which is overproduced at high light levels in photosynthesis. The xanthophylls found in the bodies of animals, and in dietary animal products, are ultimately derived from plant sources in the diet. For example, the yellow color of chicken egg yolks, fat, and skin comes from ingested xanthophylls (primarily lutein, which is often added to chicken feed for this purpose).

The yellow color of the human macula lutea (literally, *yellow spot*) in the retina of the eye results from the lutein and zeaxanthin it contains, both xanthophylls again requiring a source in the human diet to be present in the eye. These xanthophylls protect the eye from ionizing blue and ultraviolet light, which they absorb. These two specific xanthophylls do not function in the mechanism of sight, since they cannot be converted to retinal (also called retinaldehyde or vitamin A aldehyde). Their arrangement is believed to be the cause of Haidinger's brush, an entoptic phenomenon that allows to perceive the polarization of light.

Example Compounds

The group of xanthophylls includes (among many other compounds) lutein, zeaxanthin, neoxanthin, violaxanthin, flavoxanthin, and α- and β-cryptoxanthin. The latter compound is the only known **xanthophyll** to contain a beta-ionone ring, and thus β-cryptoxanthin is the only xanthophyll that is known to possess pro-vitamin A activity for mammals. Even then, it is a vitamin only for plant-eating mammals that possess the enzyme to make retinal from carotenoids that contain beta-ionone (some carnivores lack this enzyme). In species other than mammals, certain xanthophylls may be converted to hydroxylated retinal-analogues that function directly in vision. For example, with the exception of certain flies, most insects use the xanthophyll derived R-isomer of 3-hydroxyretinal for visual activities, which means that β-cryptoxanthin and other xanthophylls (such as lutein and zeaxanthin) may function as forms of visual «vitamin A» for them, while carotenes (such as beta carotene) do not.

Xanthophyll Cycle

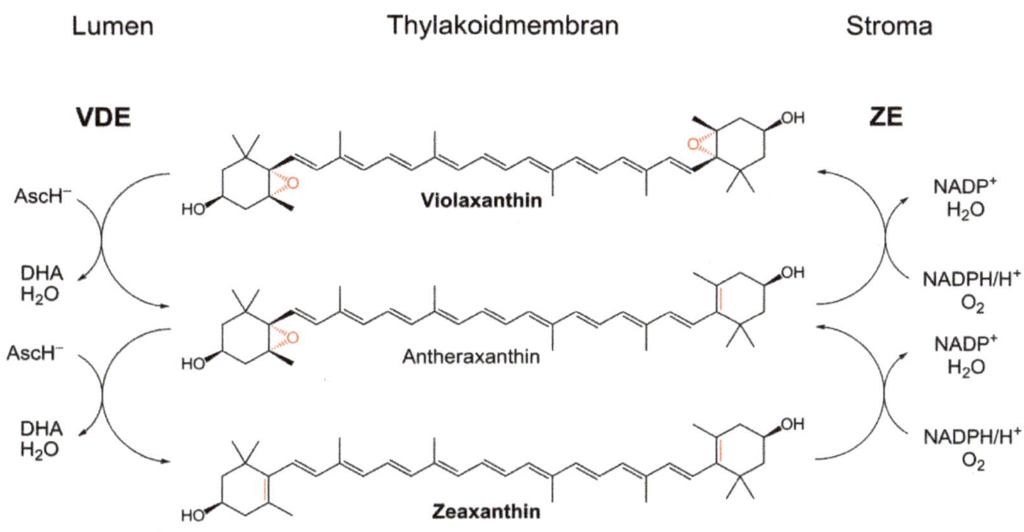

The xanthophyll cycle

The xanthophyll cycle involves the enzymatic removal of epoxy groups from xanthophylls (e.g. violaxanthin, antheraxanthin, diadinoxanthin) to create so-called de-epoxidised xanthophylls (e.g. diatoxanthin, zeaxanthin). These enzymatic cycles were found to play a key role in stimulating energy dissipation within light-harvesting antenna proteins by non-photochemical quenching- a mechanism to reduce the amount of energy that reaches the photosynthetic reaction centers. Non-photochemical quenching is one of the main ways of protecting against photoinhibition. In higher plants, there are three carotenoid pigments that are active in the xanthophyll cycle: violaxanthin, antheraxanthin, and zeaxanthin. During light stress, violaxanthin is converted to zeaxanthin via the intermediate antheraxanthin, which plays a direct photoprotective role acting as a lipid-protective anti-oxidant and by stimulating non-photochemical quenching within light-harvesting proteins. This conversion of violaxanthin to zeaxanthin is done by the enzyme violaxanthin de-epoxidase, while the reverse reaction is performed by zeaxanthin epoxidase.

In diatoms and dinoflagellates, the xanthophyll cycle consists of the pigment diadinoxanthin, which is transformed into diatoxanthin (diatoms) or dinoxanthin (dinoflagellates) under high-light conditions.

Wright et al. (Feb 2011) found that, «The increase in zeaxanthin appears to surpass the decrease in violaxanthin in spinach» and commented that the discrepancy could be explained by «a synthesis of zeaxanthin from beta-carotene», however they noted further study is required to explore this hypothesis.

Food Sources

Xanthophylls are found in all young leaves and in etiolated leaves. Examples of other rich sources include papaya, peaches, prunes, and squash, which contain lutein diesters.

Pheophytin

Pheophytin or **phaeophytin** (abbreviated **Pheo**) is a chemical compound that serves as the first electron carrier intermediate in the electron transfer pathway of Photosystem II (PS II) in plants, and the photosynthetic reaction center (RC P870) found in purple bacteria. In both PS II and RC P870, light drives electrons from the reaction center through pheophytin, which then passes the electrons to a quinone (Q_A) in RC P870 and RC P680. The overall mechanisms, roles, and purposes of the pheophytin molecules in the two transport chains are analogous to each other.

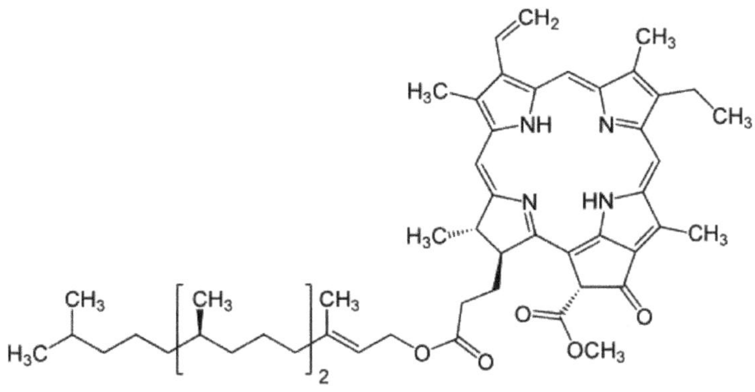

Pheophytin *a*, i.e. chlorophyll *a* without the Mg^{2+} ion.

Structure

In biochemical terms, pheophytin is a chlorophyll molecule lacking a central Mg^{2+} ion. It can be produced from chlorophyll by treatment with a weak acid, producing a dark bluish waxy pigment. The probable etymology comes from this description, with *pheo* meaning *dusky* and *phyt* meaning *vegetation*.

History and Discovery

In the 1977s, scientists Klevanik, Klimov, Shuvalov performed a series of experiments to demonstrate that it is pheophytin and not plastoquinone that serves as the primary electron acceptor in photosystem II. Using several experiments, including electron paramagnetic resonance (EPR), they were able to show that pheophytin was reducible and, therefore, the primary electron acceptor between P680 and plastoquinone (Klimov, Allakhverdiev, Klevanik, Shuvalov). This discovery was met with fierce opposition, since many believed pheophytin to be a byproduct of chlorophyll degradation. Therefore, more experiments ensued to prove that pheophytin is indeed the primary electron acceptor of PSII, occurring between P680 and plastoquinone (Klimov, Allakhverdiev, Shuvalov). The data that was obtained is as follows:

- Photo-reduction of pheophytin has been observed in various mixtures containing PSII reaction centers.
- The quantity of pheophytin is in direct proportion to the number of PSII reaction centers.
- Photo-reduction of pheophytin occurs at temperatures as low as 100K, and is observed after the reduction of plastoquinone.

These observations are all characteristic of photo-conversions of reaction center components.

Reaction in Purple Bacteria

Pheophytin is the first electron carrier intermediate in the photoreaction center (RC P870) of purple bacteria. Its involvement in this system can be broken down into 5 basic steps. The first step is excitation of the bacteriochlorophylls $(Chl)_2$ or the special pair of chlorophylls. This can be seen in the following reaction.

- $(Chl)_2$ + 1 photon → $(Chl)_2^*$ (excitation)

The second step involves the $(Chl)_2$ passing an electron to pheophytin, producing a negatively charged radical (the pheophytin) and a positively charged radical (the special pair of chlorophylls), which results in a charge separation.

- $(Chl)_2^*$ + Pheo → $\cdot(Chl)_2^+$ + $\cdot Pheo^-$ (charge separation)

The third step is the rapid electron movement to the tightly bound menaquinone, Q_A, which immediately donates the electrons to a second, loosely bound quinine (Q_B). Two electron transfers convert Q_B to its reduced form (Q_BH_2).

- $2\cdot Pheo^-$ + $2H^+$ + Q_B → 2Pheo + Q_BH_2 (quinone reduction)

The fifth and final step involves the filling of the "hole" in the special pair by an electron from a heme in cytochrome *c*. This regenerates the substrates and completes the cycle, allowing for subsequent reactions to take place.

Involvement in Photosystem II

In photosystem II, pheophytin plays a very similar role. It again acts as the first electron carrier intermediate in the photosystem. After P680 becomes excited to P680*, it transfers an electron to pheophytin, which converts the molecule into a negatively charged radical. The negatively charged pheophytin radical quickly passes its extra electron to two consecutive plastoquinone molecules. Eventually, the electrons pass through the cytochrome b_6f molecule and leaves photosystem II. The reactions outlined above in the section concerning purple bacteria give a general illustration of the actual movement of the electrons through pheophytin and the photosystem. The overall scheme is:

- Excitation
- Charge separation
- Plastoquinone reduction
- Regeneration of substrates

CHLOROPHYLL *a*

Chlorophyll a is a specific form of chlorophyll used in oxygenic photosynthesis. It absorbs most energy from wavelengths of violet-blue and orange-red light. It also reflects green/ yellow light, and as such contributes to the observed green color of most plants. This photosynthetic pigment is essential for photosynthesis in eukaryotes, cyanobacteria and prochlorophytes because of its role as primary electron donor in the electron transport chain. Chlorophyll a also transfers resonance energy in the antenna complex, ending in the reaction center where specific chlorophylls P680 and P700 are located.

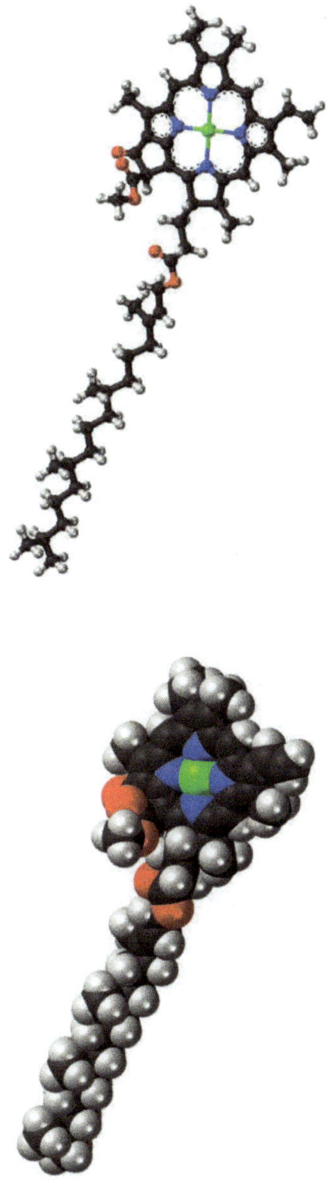

Structure of Chlorophyll a molecule showing the long hydrocarbon tail

Distribution of Chlorophyll *a*

Chlorophyll *a* is essential for most photosynthetic organisms to release chemical energy but is not the only pigment that can be used for photosynthesis. All oxygenic photosynthetic organisms use chlorophyll *a*, but differ in accessory pigments like chlorophylls *b*. Chlorophyll *a* can also be found in very small quantities in the green sulfur bacteria, an anaerobic photoautotroph. These organisms use bacteriochlorophyll and some chlorophyll *a* but do not produce oxygen. Anoxygenic photosynthesis is the term applied to this process, unlike oxygenic photosynthesis where oxygen is produced during the light reactions of photosynthesis.

Molecular Structure

The molecular structure of chlorophyll *a* consists of a chlorin ring, whose four nitrogen atoms surround a central magnesium atom, and has several other attached side chains and a hydrocarbon tail.

Chlorin Ring

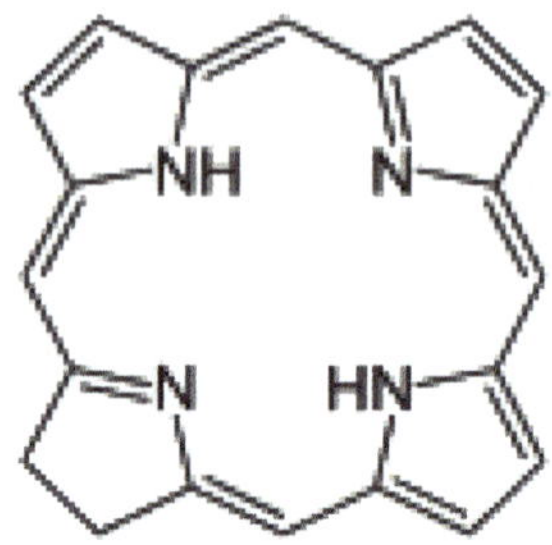

Chlorin, the central ring structure of the chlorophyll *a*

Chlorophyll *a* contains a magnesium ion encased in a large ring structure known as a chlorin. The chlorin ring is a heterocyclic compound derived from pyrrole. Four nitrogen atoms from the chlorin surround and bind the magnesium atom. The magnesium center uniquely defines the structure as a chlorophyll molecule. The porphyrin ring of bacteriochlorophyll is saturated, and lacking alternation of double and single bonds causing variation in absorption of light.

Side Chains

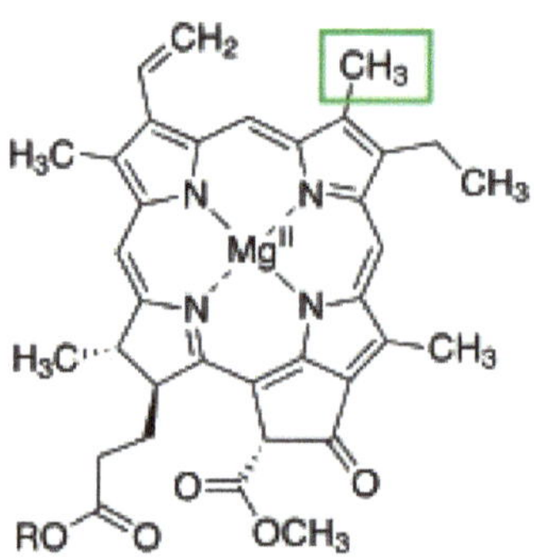

The green boxed CH_3 is the methyl group at the C-7 position chlorophyll *a*

Side chains are attached to the chlorin ring of the various chlorophyll molecules. Different side chains characterize each type of chlorophyll molecule, and alters the absorption spectrum of light. For instance, the only difference between Chlorophyll a and Chlorophyll *b* is that Chlorophyll b has an aldehyde instead of a methyl group at the C-7 position.

Hydrocarbon Tail

Chlorophyll *a* has a long hydrophobic tail, which anchors the molecule to other hydrophobic proteins in the thylakoid membrane of the chloroplast. Once detached from the porphyrin ring, this long hydrocarbon tail becomes the precursor of two biomarkers, pristane and phytane. Both of which are important in the study of geochemistry and the determination of petroleum sources.

Biosynthesis

Chlorophyll *a* biosynthetic pathway utilizes a variety of enzymes. Genes code for the enzymes on the Mg-tetrapyrroles of both bacteriochlorophyll *a* and chlorophyll *a*. It begins with glutamic acid, which is transformed into a 5-aminolevulinic acid (ALA). Two molecules of ALA are then reduced to porphobilinogen (PBG), and four molecules of PBG are then coupled, forming protoporphyrin IX. When forming protoporphyrin, Mg-chelatase acts as a catalyst for the insertion of Mg into the chlorophyll *a* structure. The pathway then uses either a light-dependent process, driven by the enzyme protochlorophyllide oxidoreductase. Protochlorophyllide is a precursor to the production of a chlorophyll *a* molecule, or a light-independent process driven by other enzymes, to form a cyclic ring, and the reduction of another ring in the structure.Attachment of the phytol tail completes the process of chlorophyll biosynthesis.

Reactions of Photosynthesis

Absorbance of Light

Light Spectrum

Chlorophyll *a* absorbs light within the violet, blue and red wavelengths while mainly reflecting green. This reflectance gives chlorophyll its green appearance. Accessory photosynthetic pigments broaden the spectrum of light absorbed, increasing the range of wavelengths that can be used in photosynthesis. The addition of chlorophyll *b* next to chlorophyll *a* extends the absorption spectrum. In low light conditions, plants produce a greater ratio of chlorophyll *b* to chlorophyll *a* molecules, increasing photosynthetic yield.

Light Gathering

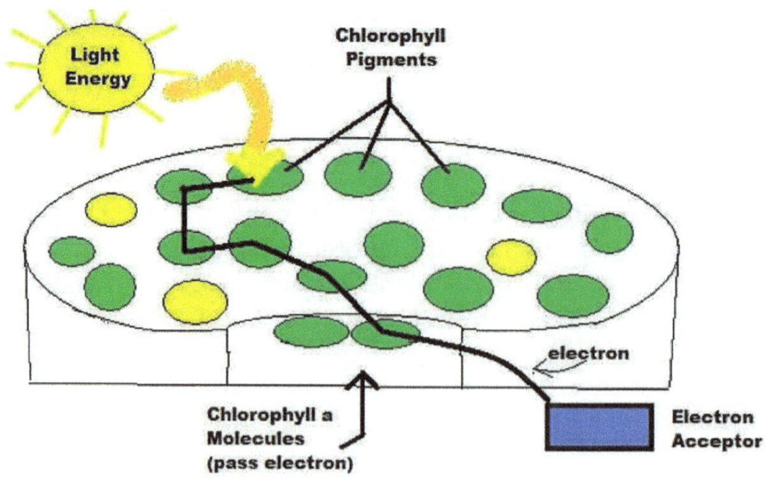

The Antenna Complex with energy transfer within the thylakoid membrane of a chloroplast. Chlorophyll *a* in the reaction center is the only pigment to pass boosted electrons to an acceptor (modified from 2).

Absorption of light by photosynthetic pigments converts photons into chemical energy. Light energy radiating onto the chloroplast strikes the pigments in the thylakoid membrane and excites their electrons. Since the chlorophyll *a* molecules only capture certain wavelengths, organisms may use accessory pigments to capture a wider range of light energy shown as the yellow circles. It then transfers captured light from one pigment to the next as resonance energy, passing energy one pigment to the other until reaching the special chlorophyll *a* molecules in the reaction center. These special chlorophyll *a* molecules are located in both photosystem II and photosystem I. They are known as P680 for Photosystem II and P700 for Photosystem I. P680 and P700 are the primary electron donors to the electron transport chain.

Primary electron Donation

Chlorophyll *a* is very important in the energy phase of photosynthesis. Two electrons need to be passed to an electron acceptor for the process of photosynthesis to proceed. Within the reaction centers of both photosystems there are a pair of chlorophyll *a* molecules that pass electrons on to the transport chain through redox reactions.

Chlorophyll *b*

Chlorophyll *b* is a form of chlorophyll. Chlorophyll *b* helps in photosynthesis by absorbing light energy. It is more soluble than chlorophyll *a* in polar solvents because of its carbonyl group. Its color is yellow, and it primarily absorbs blue light.

In land plants, the light-harvesting antennae around photosystem II contain the majority of chlorophyll *b*. Hence, in shade-adapted chloroplasts, which have an increased ratio of photosystem II to photosystem I, there is a higher ratio of chlorophyll *b* to chlorophyll *a*. This is adaptive, as increasing chlorophyll *b* increases the range of wavelengths absorbed by the shade chloroplasts.

REFERENCES

1. Mosby's Medical, Nursing and Allied Health Dictionary, Fourth Edition, Mosby-Year Book 1994, p. 273
2. "carotene". Online Etymology Dictionary.
3. Boran Altincicek; Jennifer L. Kovacs; Nicole M. Gerardo (2011). "Horizontally transferred fungal carotenoid genes in the two-spotted spider mite Tetranychus urticae". Biology Letters. 8 (2): 253–257. doi:10.1098/rsbl.2011.0704. PMC 3297373 . PMID 21920958.
4. Green AS, Tang G, Lango J, Klasing KC, Fascetti AJ (2011). "Domestic cats convert ((2) H(8))-β-carotene to ((2) H(4))-retinol following a single oral dose". Journal of Animal Physiology and Animal Nutrition. 96 (4): 681–92. doi:10.1111/j.1439-0396.2011.01196.x. PMID 21797934.
5. Linus Pauling Institute (June 2009). "Micronutrient Information Center: Carotenoids". Oregon State University. Retrieved February 15, 2010.
6. What can foods rich in beta-carotene do for you? ArchivedJune 26, 2007, at the Wayback Machine.. Whfoods.org (2004-02-26). Retrieved on 2013-03-04.
7. Pitchford, Paul (2002). Healing with Whole Foods: Asian Traditions and Modern Nutrition. North Atlantic Books. ISBN 1-55643-471-5.
8. Ajila CM, Prasada Rao UJ (2008). "Determination of carotenoids and their esters in fruits of Lycium barbarum Linnaeus by HPLC-DAD-APCI-MS". J Pharm Biomed Anal. 47 (4–5): 812–8. doi:10.1016/j.jpba.2008.04.001. PMID 18486400.
9. Olson, James Allen; Rock, Cheryl L.; Ross, A. Catharine; Underwood, Barbara A. (2006-04-23). "Dietary Supplement Fact Sheet: Vitamin A". NIH Office of Dietary Supplements website. United States Department of Agriculture. Retrieved 2007-10-26.
10. World's Healthiest Foods: Carotenoids Archived March 14, 2008, at the Wayback Machine.. Whfoods.org. Retrieved on 2013-03-04.
11. Mustard greens. WHFoods. Retrieved on 2013-03-04.
12. Simopoulos, Artemis P.; Gopalan, C., eds. (2003). Plants in Human Health and Nutrition Policy. Karger Publishers. ISBN 3-8055-7554-8.
13. Valadon, L. R. G.; Mummery, Rosemary S. "Changes in Carotenoid Composition of Certain Roses with Age". Annals of Botany. 33 (4): 671–677.
14. Adewusi, Steve R A; Bradbury, J Howard (1993). "Carotenoids in cassava: Comparison of open-column and HPLC methods of analysis". Journal of the

Science of Food and Agriculture. 62 (4): 375. doi:10.1002/jsfa.2740620411.

15. "British Cancer Organization Calls for Warning Labels on Beta-Carotene". 2000-07-31. Archived from the original on 2006-12-04. Retrieved 2007-03-15.
16. The Alpha-Tocopherol, Beta Carotene Cancer Prevention Study Group (1994). "The effect of vitamin E and beta carotene on the incidence of lung cancer and other cancers in male smokers". N Engl J Med. 330 (15): 1029–35. doi:10.1056/NEJM199404143301501. PMID 8127329.
17. Omenn GS, Goodman GE, Thornquist MD, et al. (1996). "Risk factors for lung cancer and for intervention effects in CARET, the Beta-Carotene and Retinol Efficacy Trial". J Natl Cancer Inst. 88 (21): 1550–9. doi:10.1093/jnci/88.21.1550. PMID 8901853.
18. Omenn GS, Goodman GE, Thornquist MD, et al. (1996). "Effects of a combination of beta carotene and vitamin A on lung cancer and cardiovascular disease". N Engl J Med. 334 (18): 1150–5. doi:10.1056/NEJM199605023341802. PMID 8602180.
19. Bjelakovic G, Nikolova D, Gluud LL, Simonetti RG, Gluud C (2007). "Mortality in randomized trials of antioxidant supplements for primary and secondary prevention: systematic review and meta-analysis". JAMA. 297 (8): 842–57. doi:10.1001/jama.297.8.842. PMID 17327526.
20. See the letter to JAMA by Philip Taylor and Sanford Dawsey and the reply by the authors of the original paper.
21. Grodstein F, Kang JH, Glynn RJ, Cook NR, Gaziano JM (Nov 2007). "A randomized trial of beta carotene supplementation and cognitive function in men: the Physicians' Health Study II". Arch Intern Med. 167 (20): 2184–90. doi:10.1001/archinte.167.20.2184. PMID 17998490.
22. Yanagi, Kazuhiro; Iakoubovskii, Konstantin; Kazaoui, Said; Minami, Nobutsugu; Maniwa, Yutaka; Miyata, Yasumitsu; Kataura, Hiromichi (2006). "Light-Harvesting Function of β-Carotene Inside Carbon Nanotubes" (PDF). Phys. Rev. B. 74 (15): 155420. Bibcode:2006PhRvB..74o5420Y. doi:10.1103/PhysRevB.74.155420.
23. Saito, Yuika; Yanagi, Kazuhiro; Hayazawa, Norihiko; Ishitobi, Hidekazu; Ono, Atsushi; Kataura, Hiromichi; Kawata, Satoshi (2006). "Vibrational Analysis of Organic Molecules Encapsulated in Carbon Nanotubes by Tip-Enhanced Raman Spectroscopy". Jap. J. Appl. Phys. 45 (12): 9286–9289. Bibcode:2006JaJAP..45.9286S. doi:10.1143/JJAP.45.9286.
24. Wittig G.; Pommer H.: DBP 954247, 1956
25. Wittig G.; Pommer H. (1959). Chem. Abstr. 53: 2279. Missing or empty |title=(help)
26. US patent 2609396, Inhoffen Hans Herloff & Pommer Horst, "Compounds with the carbon skeleton of beta-carotene and process for the manufacture thereof", published 1952-09-02
27. UK Food Standards Agency: "Current EU approved additives and their E Numbers". Retrieved 2011-10-27.

28. Australia New Zealand Food Standards Code"Standard 1.2.4 – Labelling of ingredients". Retrieved 2014-12-22.

29. US FDA: "Food Additive Status List". Retrieved 2014-12-22.

30. http://dictionary.reference.com/browse/pheophytin Merriam-Webster Medical Dictionary definition of Pheophytin

31. http://dictionary.reference.com/browse/pheo- Definition of pheo in The American Heritage Stedman's Medical Dictionary

32. phyt. CollinsDictionary.com. Collins English Dictionary - Complete & Unabridged 11th Edition. Retrieved October 19, 2012.

33. Anatolievich, Kiper Ruslan. "chlorophyll a". http://chemister.ru. Retrieved 2014-08-23. External link in |website= (help)

34. Lide, David R., ed. (2009). CRC Handbook of Chemistry and Physics (90th ed.). Boca Raton, Florida: CRC Press. ISBN 978-1-4200-9084-0.

35. PHOTOSYNTHESIS

36. Raven, Peter H.; Evert, Ray F.; Eichhorn, Susan E. (2005). "Photosynthesis, Light, and Life". Biology of Plants (7th ed.). W.H. Freeman. pp. 119–127. ISBN 0-7167-9811-5.

37. Papageorgiou,G, and Govindjee (2004). "Chlorophyll a Fluorescence, A Signature of Photosynthesis". 19. Springer: 14,48,86

38. Eisen JA, Nelson KE, Paulsen IT, et al. (July 2002). "The complete genome sequence of Chlorobium tepidum TLS, a photosynthetic, anaerobic, green-sulfur bacterium". Proc. Natl. Acad. Sci. U.S.A. 99 (14): 9509–14. doi:10.1073/pnas.132181499. PMC 123171 . PMID 12093901. See pages 9514,48,86.

39. Zeiger, Eduardo; Taiz, Lincoln (2006). "Ch. 7: Topic 7.11: Chlorophyll Biosynthesis". Plant physiology (4th ed.). Sunderland, Mass: Sinauer Associates. ISBN 0-87893-856-7.

40. Campbell, Mary K.; Farrell, Shawn O. (20 November 2007). Biochemistry (6th ed.). Cengage Learning. p. 647. ISBN 978-0-495-39041-1.

41. Lange, L.; Nobel, P.; Osmond, C.; Ziegler, H. (1981). Physiological Plant Ecology I – Responses to the Physical Environment. 12A. Springer-Verlag. pp. 67, 259.

42. Niedzwiedzki, Dariusz M.; Blankenship, Robert E. (December 2010). "Singlet and triplet excited state properties of natural chlorophylls and bacteriochlorophylls". Photosynthesis Research. 106 (3): 227–238. doi:10.1007/s11120-010-9598-9. PMID 21086044.

43. Suzuki JY, Bollivar DW, Bauer CE (1997). "Genetic Analysis of Chlorophyll biosynthesis." (PDF). Annu. Rev. Genet. 31 (1): 61–89. doi:10.1146/annurev.genet.31.1.61.

44. Zeiger & Taiz 2006, Figure 7.11.A: The biosynthetic pathway of chlorophyll

45. Ishikita H, Saenger W, Biesiadka J, Loll B, Knapp EW (June 2006). "How photosynthetic reaction centers control oxidation power in chlorophyll pairs

P680, P700, and P870". Proc. Natl. Acad. Sci. U.S.A. 103 (26): 9855–60. doi:10.1073/pnas.0601446103. PMC 1502543 . PMID 16788069.

46. Lide, David R., ed. (2009). CRC Handbook of Chemistry and Physics (90th ed.). Boca Raton, Florida: CRC Press. ISBN 978-1-4200-9084-0.
47. photosynthesis pigments
48. Kitajima and Hogan 2003

Index